Think In Systems

The Art of Strategic Planning, Effective Problem Solving, And Lasting Results

Zoe McKey

www.zoemckey.com
zoemckey@gmail.com

Copyright © 2019 by Zoe McKey. All rights reserved.

No part of this publication may be reproduced, stored in a retrieval system, or transmitted in any form or by any means, electronic, mechanical, photocopying, recording, scanning or otherwise, except as permitted under Section 107 or 108 of the 1976 United States Copyright Act, without the prior written permission of the author.

Limit of Liability / Disclaimer of Warranty: The author makes no representations or warranties with respect to the accuracy or completeness of the contents of this work and specifically disclaims all warranties, including without limitation warranties of fitness for a particular purpose. No warranty may be created or extended by sales or promotional materials. The advice and recipes contained herein may not be suitable for everyone. This work is sold with the understanding that the author is not engaged in rendering medical, legal or other professional advice or services. If professional assistance is required, the services of a competent professional person should be sought. The author shall not be liable for damages arising herefrom. The fact that an individual, organization of website is referred to in this work as a citation and/or potential source of further information does not mean that the author endorses the information the individual, organization to website may provide or recommendations they/it may make. Further, readers should be aware that Internet websites listed in this work might have changed or disappeared between when this work was written and when it is read.

For general information on the products and services or to obtain technical support, please contact the author.

Thank you for choosing my book! I would like to show my appreciation for the trust you gave me by giving a **FREE GIFT** for you!

For more information visit https://www.zoemckey.com.

The kit shares *10 key practices to help you to:*

- *discover your true self,*
- *find your life areas that need improvement,*
- *find self-forgiveness,*
- *become better at socializing,*
- *lead a life of gratitude and purpose.*

The kit contains extra actionable worksheets with practice exercises for deeper learning.

Visit https://www.zoemckey.com and download your Free copy now!

Table of Contents

Introduction 11

Before We Get Started… 19

Chapter 1 – The Beginning 31

Chapter 2 - Elements of Systems Thinking 53

Chapter 3 - How Do Systems Work? 79

Chapter 4 - Understanding Bottlenecks, Leverage, and Feedback Loops 107

Chapter 5 - Problems That Make the Adoption of Systems Thinking Harder 117

Chapter 6 - How to Shift to Systems Thinking . 135

Chapter 7 - Solving Everyday and Complex Problems with Systems Thinking 155

Chapter 8 - Social Problems and Systems Thinking 181

Chapter 9 - The Story of the Bins 203

Chapter 10 - Practice Systems Thinking as an Individual 219

Conclusion...235

One more moment, please.................................239

Reference..251

Endnotes 257

Introduction

It was a rainy September morning. I will never forget it. I was rushing to my first-ever university class. Puttering with my class schedule and location, I entered the modern university building. "What was the name of that class again?" I thought with a buzz of excitement only freshmen feel. "The Theory of International Relations," I read off the paper. I had zero thoughts about what this class would end up being. I understood each of the words individually, but put together, they lost all meaning.

I arrived just before the professor. Because of my delayed arrival, I only found available seating in the first row. I felt exposed to the sharp eyes of the professor, who arrived just when I was trying to fish out my textbook, but from the sudden shock of her appearance, I lost control of my shaking hands, pouring the contents of my backpack onto the ground.

"What's your name?" she asked me.

"Z-Z-Zoe?" I asked her back, which she rewarded with a stern glance. "I mean, I am Zoe, yes," I repeated myself, trying to collect my things and the little pride I had left. Great, now she would remember me until graduation.

"Well, Zoe, you just gave me a great opportunity to explain the topic of the class in an easy manner. Thank you." I was so surprised that I forgot to say anything and just nodded. She then turned to the class. "What happened to Zoe?" she asked everyone.

After a few seconds of confused, simultaneous mumbling, someone said, "She dropped her bag." The professor nodded and gave the class a look that obviously meant she wasn't satisfied with the answer. So someone else added, "She dropped the bag off the table because you scared her."

All the blood rushed out of my face when I saw the professor's eyes bulge for a second, but then she laughed. "Yes, that is a better description of the event." Then, without waiting for another creative explanation about my shame, she continued.

"In linear thinking terms, you are correct. She got surprised, she lost balance, and she dropped her bag. But there is so much more to it. Why did the bag fall to the ground versus just floating around? Why did she choose to take out her book at the last minute? Was she late? Why was she late? Did her bus arrive late? Why did the bus arrive late?

Was that the fault of a traffic jam, or some internal organizational problem at the bus company? You see, there is so much more to a dropped bag than we think." She totally got our attention.

"As you're now starting to understand, there is much more to a simple event than what it seems to be on the surface. The theory of international relations aims to give a conceptual framework upon which complex relations between the actors of a system can be analyzed."

Here, she lost us... Conceptual what? She saw that, and with a chuckle, she explained herself.

"In other words, we will take a look at the relationship some countries—actors—have with each other, and within. We will learn different theories through which we can explain international phenomena like why wars happen, what the challenges of war preventive policies are, and why we can't truly eradicate hunger, drug wars, and other issues from the world." Our attention was captured again.

"To be able to discuss these topics meaningfully and understand each other, first I need to teach you to shift your thinking patterns. There is an entire discipline built upon complex thinking that goes way beyond the regular cause-effect thinking. This is systems thinking."

And there it was, my first encounter with the concept of systems thinking, at the age of 19, in the very first class I had at the university. Ever since then, it has been quite a journey learning about systems thinking. With a decade of active and profound learning about it, I overcame my hesitation and decided to collect my experience and understanding into a book. Systems thinking had life-changing benefits for my life early on, and I'm very grateful for it. This is not as popular or widespread of a topic as it should be, partially because it is a new discipline and partially because it operates with professional jargon that can be hard to grasp. My goal with this book is to deliver this quintessential discipline to you in an understandable language, profound enough to improve your life and relationships, understand world events, and find better solutions to your problems.

You won't become a graduate systems analyst by reading this book. If that's your goal, I can recommend some more advanced books on systems thinking. This book is an introduction to systems thinking, very handy for those people who are learning about it from scratch, or for those who wish to recapitulate the basic essentials. This book will present the theory of systems thinking in the first few chapters and give you practical examples and "homework" in the latter chapters. After reading this book, you'll be able to understand the jargon of the basics of systems thinking, design simple loop diagrams, behavior-over-time diagrams, and more. I also provide plenty of references and resources as we go to help you advance your knowledge on the topic, if you wish.

And if you were wondering, the professor didn't forget my name until graduation. But more importantly, I'll never forget her easy-to-understand explanation of what international relations theory, and within that, systems thinking, is. Her clever way of simplifying the complex made me interested in this topic in the first place. By no means will I be as great a teacher as my professor was, but I will simplify and amplify your knowledge on systems thinking to the best of my ability.

Before We Get Started...

I would like to introduce you to the great minds who contributed to the foundation of systems thinking as a discipline.

The thinker who is considered to be one of the founders of systems thinking—or as he called it, general systems theory (GST)—to my great pride is an Austro-Hungarian biologist, Ludwig von Bertalanffy. His explanation of general systems theory was interdisciplinary, rather than a standalone discipline. His practice described how the components of a system interact with each other—unsurprisingly, in biology or cybernetics. In 1934, he came up with his groundbreaking mathematical model of an organism's growth over time, which is still applied today. He was the father of open systems, which I will talk about later in this book.[i]

His general systems theory also helped provide alternatives to conventional organization, putting the emphasis on a holistic approach versus a reductionist approach and organisms versus mechanisms.[ii]

The systems thinking we use today evolved from the general systems theory. Systems thinking admits that systems have a certain dynamic that is subjected to external factors like feedback mechanisms, delays, and other aspects. The dynamic of a system can often be counterintuitive. For example, everybody works hard to make the country's economy flourish, yet there is a constant economic decline. Why? Systems mapping and systems dynamics modeling can answer this question and help us understand the system's behavior over time. By understanding the behavior, we can find the causes as to why the economy is not performing well despite everybody's best effort. Then we can inject a targeted change to improve the system's future behavior and achieve its goal.

Conventional, reductionist thinking aims to solve everyday, linear problems. For example, if we run out of milk, we need to buy more milk. If the car runs out of gas, we need to fill it up. If our stomach is growling, we need to eat. These are easy-to-fix problems that don't require a meta-analysis to be fixed, right?

What about the non-linear problems that can't be fixed with a cause-effect analysis? What is a non-linear problem? For example, predator-prey relationships. Why are rabbits overpopulating? Why are there more crop-eating insects one year after an effective cleansing with pesticides? Why do certain species go extinct? Market reactions are also non-linear. Why is the market reacting better or worse to a new product in certain months?

Complex socioeconomic issues like legalizing marijuana, gay marriage, or gun control laws are also non-linear. A simple "just because this happened, we will do that" kind of thinking, or worse, applied problem-solving attempt, will produce serious consequences. Any change in these matters affects many actors differently. The benefit of one actor may be the loss of another actor. How can you find a common ground with everybody? Systems thinking seeks the solution to these kinds of problems.

Do you feel overwhelmed now, thinking, "Okay, this sounds profound, but I don't really care about rabbits, pesticides, and drug wars. I just want to fix my own life"? You are right. Learning to think in systems will help you understand all these global issues, but that doesn't mean you can't easily transfer its principles to explain and seek solutions to your personal problems. For example, if you run out of milk, you can go and buy another bottle. This action requires only linear, cause-effect thinking. But if you don't have the money or time to buy another bottle, that's another issue. You can't solve this issue by simply saying "I don't have money because I'm dumb and I can't find a better-paying job." or "I have so much to work on that I don't have time to buy milk." These problems require a more complex and deeper analysis to find a proper solution. We tend to treat the symptoms of our problems, like borrowing money for milk or taking a day off work as sick leave, but in the long run, these temporary fixes won't make the problem go away.

Barry Richmond in his book *An Introduction to Systems Thinking* defines systems thinking as "the art and science of making reliable inferences about behavior by developing an increasingly deep understanding of the underlying structure."[iii]

Weinberg, the author of *An Introduction of General Systems Thinking*, asks three systems thinking questions that could be a good foundation for understanding any issue in any situation. These questions are:

1. Why do I see what I see?
2. Why do things stay the same?
3. Why do things change?[iv]

In his book, he answers these questions with interesting and thought-provoking examples like the unintended consequences of waste heat from nuclear reactors, or of targeted pesticides (which I will talk about in detail later in this book).

Let's practice systems thinking unknowingly a little bit. Choose an event from your life.

Anything. Now try to answer the three questions above in the light of your chosen event. I chose the following example: The more I travel, the less connection I feel with my friends at home. I feel that while I change a lot, nothing changes at home.

1. **Why do I see (or feel) what I see (or feel)?**

   Because I experience very different things than they do, thus there are fewer common topics for us to connect on.

2. **Why do things stay the same?**

   Even though we don't share the same experiences, we can still connect on other levels like nostalgia, common acquaintances, and feelings beyond what a traveling lifestyle affects. After all, regardless if we live a settled or nomadic lifestyle, there are human aspects that evolve over time that can connect people. It is incorrect to assume they didn't change at all and there is nothing left to talk about. Their change simply doesn't point to the direction of my change. It may have a different dynamic and different priorities.

3. **Why do things change?**

Things change because by experiencing different things and getting to know more people, I feel I become more understanding, I can accept other cultures better, and I harbor much fewer biases than before. My friends, on the other hand, don't experience these things, and therefore there is not only a discrepancy in shared experience, but also in worldview. These two differences have an accumulated effect in making us uninterested, or even annoyed with each other.

As you can see, just by having this quick little brainstorming session, I could articulate the problem more accurately than I could just by saying, "My old friends are boring, whatever." Thanks to these answers, I also found a solution to the problem—which points I should focus on to connect with my old friends. This is a very simple case, of course.

To solve problems long term, you need to focus on three things, according to Balle, the author of *Managing with Systems Thinking: Making Dynamics Work for You in Business Decision Making*.

1. Detect patterns, not just events.
2. Use circular causality (feedback loops, which I will present later in this book).

3. Focus on the relationships, rather than the parts.[v]

Balle focuses on applying systems thinking principles primarily to the workplace. He recognized that typical management solutions are usually treating symptoms, and thus "solved" problems reoccur in the long run. This thesis can be applied to anything, not just mismanagement in organizations. What are the reoccurring problems in your life? Try to think about them from the perspective of Balle's three points.

The real goal of systems thinking as a discipline is to help you understand and address system problems. As a consequence, you'll be able to investigate real world problems just as much as your own with a more comprehensive approach and find better solutions. To start, observe models and systems at work in your own life. When and how do they start, where would they end, and how could you improve them?

How to do all this? The following chapters will finally tell you.

Chapter 1 – The Beginning

If I tell you to think about your life, you will probably come up with a picture in your mind of yourself. When I ask you to think about the things you do, suddenly the image in your mind becomes connected to your laptop, your toothbrush, coffee mug, a few people to whom you speak daily, and so on. Is that it? Is that your life? Not really.

Your picture is still not complete. You have myriad other things going on apart from your routine. Let's put a little more effort into this imagination game. What else is there? I can see you are focusing a lot right now. Your picture becomes a lot more intense. In the middle of your life, you see yourself, then you have your toothbrush, your cellphone, your computer, your boss, the different people you talk to, the news you read, the food you eat, maybe your bed, some cleaning tools, your kids, and much more. Now you can see activities, people, and events connected to your boss too. To the newspaper, you realize you can connect journalists, events, and an institution from where it operates. The toothbrush was made by workers, developed by dental specialists, produced in China (probably, like everything else). The list could go on and on and on.

I bet by the time you finished reading this paragraph, you felt like you were watching yourself from an airplane. First you had your focus on yourself and then you zoomed out, inviting in the broader perspective of the system you are operating in. Because that's what this exercise was all about—imagining yourself in this infinitely complex system called life.

Our life is a system, yet we don't think of it this way. Thinking of our lives as a system that can be organized and broken down into and analyzed as smaller concepts helps us focus and improve a particular issue in isolation. However, we can also see that just because we improve an issue in isolation it won't necessarily make the entire system better. For example, if there is famine in your country, the system you live in won't get better just because you alone filled your belly. But if everyone found a way to fill their bellies, or if you found a way to fill everyone's belly every day, that personal change would affect the system. This analytical approach is helpful in understanding how everything in our lives is interconnected—how one event influences another, and simultaneously, what impact it has on the system as a whole.

The system of your life has a lot of moving parts. While you have your physical possessions—your clothes, house, and other objects—you also have abstract parts, like your beliefs, identity, and values. To make things more complicated, you also have components you don't immediately have control of, like your relationships, the economy you live in, finances, and health.

You have all these things that make up your life, but to further complicate things, they influence each other! For example, do you think your finances impact your health? One might say that those with more money can afford a gym membership and healthier food. Lower socioeconomic classes might only be able to afford fast food and processed foods that aren't healthy. And maybe your looks affect your social relationships. If only you were better-looking, or if only you weren't *so* good-looking, then people would take you more seriously.

These interactions between parts in your life are all part of your life's system. So, systems thinking is a set of analysis tools and techniques that aim to look at the complex parts of your life and break them down into something that's a bit easier to see, digest, and ultimately improve.

Your life isn't the only system you're used to. In fact, many things can be considered systems. There are open and closed systems. Books are closed systems, because while they are sitting on your shelf, they have no inputs or outputs. On the other hand, open systems are pretty much everything around you. The pipe system in your house, jungles, and even your car are all open systems. They have set boundaries within which they operate. They also have a number of inputs and outputs.

Just like the jungle and the pipe system, your life is an open system. You get inputs, like money, a can of milk, or a gift, and you provide outputs like labor, the sharing of personal experiences, giving a gift, or engaging in different activities.

Life is a little more complicated than an open system because its boundaries are not set. Why? Because we humans are also part of the evolutionary system, which changes its boundaries constantly. We can look at our lives in the following way: Our life is an open system, on one hand. We have certain sets of boundaries we operate in—financial boundaries, health boundaries, boundaries posed by norms, ethics, laws, etc. On the other hand, our life is part of a bigger system called an evolutionary system, where we are exposed to the constant changes of evolution we can't influence. For example, the way we acclimated to new environments was by learning to use tools, and thus we became better adapted to operating within larger communities (like big cities versus a small tribe).

While our lives can't be completely broken down into a set of diagrams, the diagrams can help us better understand the complex ideas and events in the world that surrounds us. Every system wants to work well. It is the underlying idea of designing them: functionality. Using systems to understand your life can actually help you get to know yourself better.

The system is a model.

Systems are like maps. They model a situation, but they are not the situation itself, just like a map is not the land itself.

Systems are simplified models of reality. Their function is not to present the world as it is, but to make it more understandable and easier to correct or improve. Most systems are already "invented." They have a basic functioning structure. You can change the inputs to a system and make them deliver the output you desire.

Let's look at a practical example: making scrambled eggs. The system behind this culinary delicacy is the recipe. You need eggs, salt, a frying pan, oil or cooking spray (if you don't have a nonstick pan), and fire. These are the elements of the system model named scrambled eggs. (Picture 1) You don't have to invent the system of how to make scrambled eggs. Someone else did it before you. All you have to do is use this system, mix the ingredients together (input), and eat the food (output). (Picture 2)

A system can (and often should) be improved. In our case, we can add our favorite ingredients to make the scrambled eggs more appealing to our taste buds. I would add some pepper and cheese, and some vegetables like tomatoes, green peppers, and onions to turn my plain scrambled eggs into a royal meal. (Picture 3)

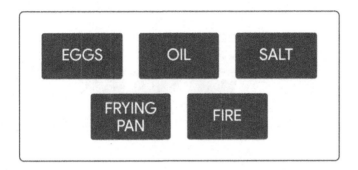

Picture 1: The mode of a simplified system

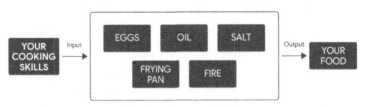

Picture 2: The effect on the simplified system of scrambled eggs.

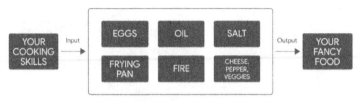

THE IMPROVED SYSTEM OF SCRAMBLED EGGS

Picture 3: The improved system of scrambled eggs

You could illustrate on a diagram the process of brushing your teeth as well. First, you have your unbrushed teeth. Then you need the toothbrush, toothpaste, and the brushing motion (the system of tooth brushing). The output would be clean teeth. If you wanted to improve this system, you could add flossing and mouthwash to your routine, and so on. Our daily lives are full of little systems like these.

But the question of peace and war, atomic armament or disarmament, healthcare benefit increases or cuts, tax cuts, and environmental issues all get decided more or less on the same basis—with the help of slightly more complex system diagrams than our scrambled egg one. I'll talk more about complex systems later in the book.

Systems—and thinking in them—surround us, and yet it's not the most natural way to think on a daily basis. We don't go around overanalyzing the creation of scrambled eggs. And we shouldn't. It would make us crazy, because every single aspect of life can be viewed with a system lens. It is, however, a good practice to try to deconstruct easy-to-follow parts of our lives into systems for practice, because honing our ability to notice easy connections in life will help us see more compounded events as we progress.

We may find better solutions for workplace or relationship issues just by taking a step back, leaving the ground of emotions and cause-effect thinking, and starting an objective analysis. Not to mention the benefits we get in terms of better understanding local and global politics, economics, and decision-making. Everything is interconnected with everything. You are connected with your local government, which is beholden to the national government. Your country is part of at least one international organization. Each element has an effect on the other.

The good news for you as an individual is that you can borrow the well-designed systems of the world and use them in your life. Let's take the system of supply and demand from economics. The basis of this system or model says that for any product in a market, there is an equilibrium point where people want (demand) and provide (supply) the same amount of the product. As we can see in Picture 4, there is a point where there is equilibrium between what a product costs based on how many people want to buy it. If the demand grew for the product (demand's line moves right or upwards), the price would rise. Why? Because the product wouldn't be available for everybody who wants to buy it, and thus people would pay a higher price to get it.

The opposite is also true. If the demand drops, prices go lower to tempt people into buying the product again. It makes sense, right? Based on Adam Smith's free market theory, when the demand rises, so do the prices, up to a point where people wouldn't pay the price requested for the product, so demand would drop, and eventually it would establish a new equilibrium. Or if demand drops, prices drop. People would become more interested in buying the product for a lower price. Thus the demand would rise again and create a new equilibrium.

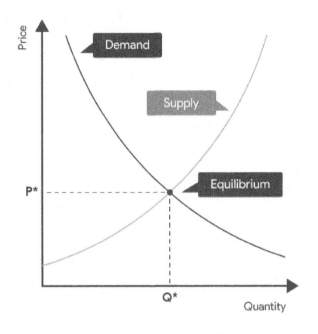

Picture 4: The supply and demand graph

Let's apply this simple system to your job-hunting struggles. You are an actor and desperately want to play Romeo in the next season at your local theater. The problem is that there are 99 other Romeos queuing for the role. What does this mean? It means the demand for the role is high. If the Romeo market were in equilibrium, there would be 100 roles available for Romeos (bigger supply), or just one person applying for the job (lower demand). In our case, there is a low supply (one role) and high demand (100 Romeos). On a diagram, it would look like this:

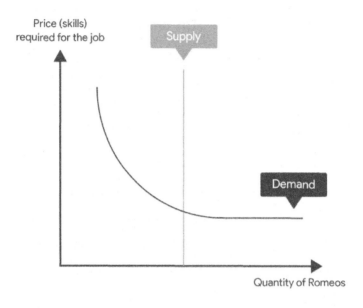

Picture 5: A one-Romeo market place

In practice, this means that you are one of a hundred Romeos auditioning for the part, and the chances of getting selected are fairly low. The market is distorted in favor of the supply (or supplier) who can thus set a higher price (better skills, more experience, best-looking actor, etc.).

No worries, Romeo. The process of getting a job is also a system, and thus it can be improved. In our case, practicing your acting skills, getting in shape, or making a profound Shakespeare background check can all increase your chances of selection. It can also increase your chances if you give your application to multiple theaters with *Romeo & Juliet* on their schedule (increasing the supply). The demand still overpowers the supply, but now you have much higher chances than before.

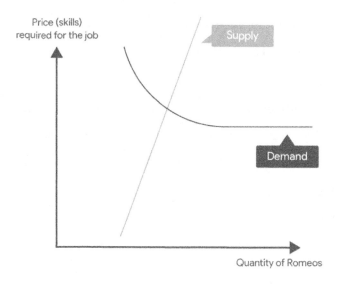

Picture 6: A multiple-Romeo market place

As you can see, with a basic understanding of systems thinking and applying a simple system that already works, you can work out new, better strategies. You can also take the existing system to understand a different area of life with it (like we did with the supply and demand model and job hunting). Then you can modify it to better suit your needs. Giving your personal input, you can modify the system's structure, be it scrambled eggs, the chance to nail a job, or even becoming president of a country.

Some of us are already systems thinkers without realizing it. Are you the type of person who, when you wake up, makes a checklist of the 10 most important things you need to do that day? Then you start working on task number one, then two, until you reach 10? This is your system, then. If you are a real systems thinker, you try to go beyond this number and list out 30-50 tasks—not only for the day, but for the week—prioritize them in a logical order, analyze which task will help out the completion of another task, etc.

Some—many—people are not like this. They wake up in the morning asking themselves "What should I do today?" and then they start doing something without having a clear purpose or clear priorities. This is a system too. But chances are slimmer that this system will go through improvement over time.

The good news is, even if you are not a list-maker, the big system of your life can be broken down into smaller sub-systems, and they can be improved bottom-up. You might think that life just happens to you on a daily basis, but I will help you deconstruct it to have a say in it.

Using systems thinking can help you look at the patterns of your life through a different perspective. Putting the emphasis on the relationship between things and events in your life, rather than the things themselves, helps you to see your current position and the long-term perspective more clearly. I will help you to slowly but surely systemize your life. It's like planting a seed. When that seed finally sprouts, you are only seeing a small part of the plant. The majority of it is left beneath the surface. A lot of your life isn't seen on the surface, and you have to look deeper to find out what is really going on.

Systems thinking will change the way you analyze your life, and this book is going to help you do that. Throughout, we will discuss the basics of systems thinking. Everything you need to know to practice this way of thinking will be presented to you. Plus, I'll go over the three biggest errors that happen within systems. Once we cover all of that, I'll help you discover how to change your life and relationships to become exactly what you hope for through systems thinking.

Chapter 2 - Elements of Systems Thinking

The Parts of a System[vi]

Essentially, analysts differentiate three major components of systems. For those nonhuman systems, you'll have:
- Elements;
- interconnections; and
- function.

For human systems, you'll have:
- elements;
- interconnections; and
- purpose.

Elements

The elements are the "actors" in a system. Actors are the official name for any active or passive partaker in systems thinking-related disciplines. Thus, from now on when I talk about actors, I don't mean Brad Pitt and Danny DeVito. This being said, not only humans can be actors. For example, if you took a look at your nervous system, you'd find that the elements are your brain, spinal cord, nerve fibers, skeletal muscles, and internal organs. These are actors too. The elements or actors are all interconnected due to the chemical reactions sent through your spinal cord and nerve fibers by your brain. The function of your nervous system is to transmit different signals throughout your body. It's like your mainframe of your body, and it is a system.

A basketball team is also a system. The elements consist of the players, the coaches and staff, and the basketball. Human and nonhuman actors. This system is connected by the rules of the game, the communications that occur between the coaches and the players, and the laws of physics. Since basketball as a sport is a human system, it has a purpose—which in most cases is winning and continuing on to the championships. But it can also be a friendly basketball match between coworkers on an active Saturday afternoon. Here, the purpose gets blurred. The players still want to win, but they also want to have fun, be more active, and strengthen friendships. Optimally, in such kinds of meet-ups, there are no hard feelings as can often happen during national championships where millions of dollars are at stake.

The easiest thing to pick out of a system is its elements, primarily because elements are made up of things that you can see or touch. The elements that make up a flower, for example, are going to be the roots, the stem, the leaves, the seeds, and the petals. The elements of a school are the students, teachers, learning materials, education laws, school curriculum, the building, the classrooms, and even the chalk.

There are also non-physical elements in a system. You can't touch a company's loyalty, nor can you touch the excellence of the best surgeon at a hospital, or the pride belonging to the team that wins the basketball championship. These are still elements of these systems. Elements could continue on and on within a system. Each element could probably be divided into even more elements, but we'll save that for a rainy day.

Interconnections

The interconnections are usually a bit more difficult to determine, but they're everything that brings the function of a system to life. In the flower system that I described above, the interconnections would be the physical and chemical reactions that keep the flower alive, like water maintenance. If it is dry and hot outside and the petals and leaves lose water, they will give a signal to the roots to take in more water from the soil. If the soil is dry, the flower will shrink its pores on the leaves and petals to not lose too much water. This all happens through chemical reactions and the environment around them. Flowers know when they need to get more water and when they need less due to things like a hot day versus a rainy day. They are photosynthesizing. This enables them to convert light energy into chemical energy. Thus, they can fuel the organisms' activities—like signals on rainy days or drought.

Some of the interconnections you see in systems are going to be actual, physical events, like the water being drawn up from the roots of a flower, or the minimum requirements to pass a course.

However, many interconnections are going to be flows of information and signals. Because of this, they are harder to see, but you can still find them if you look hard enough.

While they are harder to detect, these are more common in human systems. It's like when you look up on Facebook which of your superiors has the most common attributives as you to increase your chances of getting your project accepted, or deciding between an expensive shoe and an imitation of it based on how much money you make, or the local government collecting data on what to invest in, improving the roads or refurbishing two parks, based on voter complaints and dissatisfaction. This flow of information is an interconnection that affects decision-making, but it's not necessarily a tangible thing you can see or touch.

Purpose or function

When looking at a system, they all have a purpose (human systems) or a function (nonhuman systems). This is the least obvious part of a system. It's not expressed on a billboard, but it's still there. By watching what a system does, you can often find its purpose or function. If you see a woman running up a set of stairs, running down a set of stairs, and running back up once again, what is her purpose? She probably isn't trying to visit someone who lives on the top floor, but rather get an exercise or workout. If a man claims that his main goal is to spend more time with his family, but then he prioritizes his work affairs all the time and goes out on the weekend with friends, then family is clearly not his priority. You can deduce purpose or function from actions and behavior, not words or goals.

Concepts of Systems Thinking[vii]

As we go through systems thinking in this book, I will highlight the main concepts you need to know. I'll explain easy systems thinking models more in-depth as we continue. There will be a practical part too, but to get there—and for you to properly understand and use them—first you need to grind through a short theoretical part. These concepts will help deepen your knowledge and understanding of the process of systems thinking. Once you know all the necessary explanations, you'll find the practical part easy to follow.

Interconnectedness

Most of us are used to thinking in a linear way, but systems thinking operates in a rather circular way of thinking. Why? Because everything is connected. Remember the exercise you did in the beginning of this book, imagining your life and then zooming out? Just like humans rely on nutrition, oxygen, and water to stay alive, plants rely on the sun and carbon dioxide. What is happening here? Plants "inhale" carbon dioxide and "exhale" oxygen. Human do it in reverse. Everything that exists in the world relies on something else to keep existing, right?

Even if you think about inanimate objects, they are still relying on something for their existence. For example, your laptop needs power, so it relies on electricity. Scrambled eggs rely on a chicken to lay eggs. A table relies on a tree and a carpenter, or a furniture factory, to be made. This is what it means for everything to be interconnected. It's not a structured relationship between the living and non-living. Everything you see on a daily basis is somehow connected to something else, which somehow connects to you. Systems thinking, in a sense, presents the fundamental principle of life—that everything in this world is interconnected; the actors affect each other directly or indirectly. As they say, a "butterfly flapping its wings in Brazil could set off a tornado in Texas."

Donella Meadows put interconnectedness in a more explicit explanation: "A system is a set of related components that work together in a particular environment to perform whatever functions are required to achieve the systems objective."

Synthesis

Systems thinking might give the impression of being the discipline of profound analysis. In fact, its goal is quite the opposite, since analysis is about breaking something into smaller, digestible components to make it easier to examine. But this goes against the aim of systems thinking, which is deemed to look at things in their complex nature. So what is synthesis, then? In conventional terms, synthesis means "the combination of two or more elements to create something new."

Systems are complex, interconnected, dynamic phenomena. A reductionist approach won't give justice to the explanation of happenings within a system. A more holistic approach is needed to explain systems dynamics. You need to understand the different elements, connections, functions or purposes, and the big picture of the system at the same time.

Feedback loops

I will delve deeper into feedback loops for an entire chapter later because they are essential in understanding diagrams and models. Here, I will just present them briefly. We've established that everything is interconnected; therefore, the actors have an effect on each other. These effects create certain feedback loops. To put it very simply, what goes around comes around.

For example, if you act meanly toward your spouse, they will act meanly too, which may tick you off and make you act meanly again, and so on. This type of feedback loop is called a *reinforcing feedback loop,* because the action of one actor and the reaction of the other actor reinforces and encourages the same dynamic. Other examples could be overpopulation or a birth rate decrease, getting indebted to the bank, or the spreading of a disease.

The nature of the ecosystem is to balance things out: under regular circumstances, there is enough prey for predators, enough food to feed humankind, and enough oxygen for everybody. Under natural circumstances, if prey animals—let's say rabbits—overpopulate because there wasn't a drought that year and grass was plentiful, foxes will grow in stock as well, since there will be more rabbits to feed their offspring. The supply-demand model in equilibrium works based on this feedback system too, which is called a *balancing feedback loop*.

However, if you artificially interfere with the laws of nature, you create an imbalance. For example, if you hunt down too many foxes for their fur, the stock of rabbits will grow in extreme measures, creating a reinforcing feedback loop. The spread of medication among other factors (which I'm not saying is a bad thing) made overpopulation happen—again, a reinforcing feedback loop.

It is crucial to observe and understand the type and dynamics of a feedback loop so we can make the necessary corrections where needed.

Causality

You often hear the saying that correlation does not equal causation. What this means is that just because two things are related doesn't mean that one thing is causing the other. On the other hand, causality is deemed to identify how and why one thing happens because of another thing.

This concept is about how different things influence one another within a system. Finding causality within a system helps you better understand the dynamics of feedback loops, connections, and actor dynamics—all the essential parts of a system.

Picture 7: The loop of causality

Emergence

Big things are the culmination of many small things put together. In a biological sense, we can think of the example of how life emerges from the combination of two cells. But let's look at a less abstract example: a cake. It comes from flour, sugar, eggs, milk, baking soda, heat, and so on. Emergence is the natural way things come together. When we talk about emergence, we are talking about the outcome that happens when different items interact with each other.

Like with cake, it doesn't emerge until all the ingredients fuse together and are baked at the correct temperature. It's like how an icicle forms when it's cold enough and a drop of water freezes, then another and another. These things happen when different environmental components get together and interact. Take out the flour of a cake or the water from an icicle and you don't end up with anything remotely similar to a cake or icicle. Emergence is often the most difficult part to detect and understand, but once you understand it, you'll be able to think about and predict outcomes from seemingly unrelated things. For example, when you see a tadpole, there is nothing about it that would indicate that it will turn into a frog soon enough.

Systems mapping

There are a lot of different ways you can choose to map out a system, but underneath it all, it's still the same. You have to find the elements within a system and map them out with a chosen model so you can understand their interconnections and relationships within the system. Once you've figured out what's happening in the system, you can also find what's wrong in that system and come up with solutions to improve it.

Here are a few system-mapping methods, which I will talk about in detail later in the book. We talked about some already. Picture VIII, our Romeo career diagram, is a simple "behavior over a given event (time, or price, or quantity)" diagram. There are other types of system mapping models, like the iceberg model (Picture IX), simplified loop diagrams (Picture X), and others.

The Iceberg Model

The iceberg model is best presented in Kim Daniel's book *Introduction to Systems Thinking*. The model basically argues that in a system, reoccurring events define patterns. These patterns are caused by the structure of the system. If from no other source than the movie *Titanic*, we know that icebergs don't reveal themselves 100%. Usually, only 10% of their surface is visible; the rest is underwater. But that 90% which is invisible in fact creates the iceberg's behavior at its tip—in other words, the visible action.

As you can see in Picture IX, there are four levels of the iceberg.

1. The event level

This is the proverbial tip of the iceberg, the visible part which is perceived by everyone. For example, you might wake up one morning and see riots on the streets. While this issue seemingly would have an easy solution—just dissolve the riot or wait until they clear the streets—this attitude wouldn't solve the issue behind the riot. Problems observed at the event level are the consequence of something deeper, and the iceberg model encourages us to dig deeper instead of seeking adjustment at the event level.

2. The pattern level

If we take the time and energy to look beyond the event level, we can notice some patterns. Similar, predictive events were happening in the past—for example, big social media initiatives to demand changes, smaller protests, newspaper articles, street talk about the topic of the riot, etc. By observing patterns, we can forecast and preempt issues from happening.

3. The structure level

After we have identified some patterns that reoccur over time, we need to find what is causing or triggering the pattern we found. The explanation, in most cases, is some kind of structure—for example, the continuous blind eyes and deaf ears of policy makers, a country's worst overall financial status, or an increased feeling of marginalization or hopelessness in the population could all build up to a riot. Professor John Gerber at the University of Massachusetts identifies the following four structures:

"1. Physical things—like vending machines, roads, traffic lights, or terrain.

2. Organizations—like corporations, governments, and schools.

3. Policies—like laws, regulations, and tax structures.

4. Rituals—habitual behaviors so ingrained that they are not conscious."[viii]

4. The mental model level

The last part, the bottom of the iceberg, is mental models. This is the level of attitudes, morals, beliefs, values, and expectations that enforce and maintain structures to function at their current stage. What beliefs keep the system in place? For example, beliefs and traditions we learn at home will dictate our decisions later in life, and those will be the beliefs and traditions we'll instill in our children. These are not necessarily conscious decisions. Mental models that could affect the outbreak of the riots are: the belief of the current zeitgeist being in opposition with the legislation, the despair of not being able to afford certain things, offended or ignored national, racial, or religious beliefs, and so on.

Exercise:

Create your own iceberg model analysis! Think about something that has happened with you (or in the world) in the past days or weeks. It can be a political controversy, an economic issue, a natural disaster, or a personal struggle. Do you have it? Now, based on the four stages of the iceberg model, describe the stages of your event. What was observable about the event? What were the patterns? What structure triggered the patterns? What mental models fueled the structure?

If you want to delve extra deep into this exercise, choose another event which is similar in nature and do the same exercise, but now upside down. First, try to find the mental models behind the event, then the structure, then the patterns, and finally conclude how all of this shows itself at the event level.

Your tasks are not done yet. Now that you went through (hopefully) two iceberg analyses, try to define for yourself how this practice broadened your perspective. If you got some perspective, is it relevant? Helpful? Can you build a solution from the new knowledge? Did you find the actual part that needs intervention? How would you intervene to improve the system? How would the system improve by that intervention?

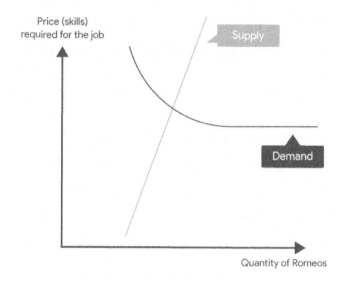

Picture 8: Behavior over price/quantity diagram

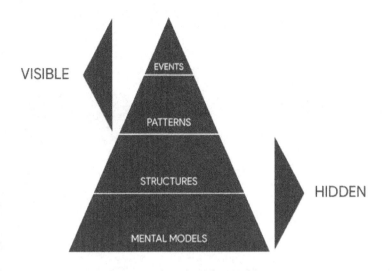

Picture 9:ix *The Iceberg Model.*

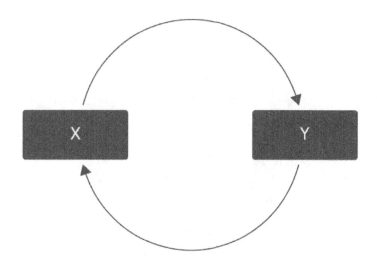

SIMPLIFIED LOOP DIAGRAM

Picture 10: *Simplified model of understanding feedback loops.*

Chapter 3 - How Do Systems Work?

The Most Important Part of a System

Did you wonder which part might be the most important of a system when you were reading about them? If not, don't worry. I will tell you soon. Let's recap the three parts of a system: it has elements, interconnections, and a function or purpose. To find the most important part of a system, picture a random system and imagine its parts changed one by one. First, take your system and change its elements mentally, then its interconnections, and finally, its purpose.

Let's look at this exercise in detail. I picked a school and its students as my system.

Generally speaking, when you change elements, there isn't a huge effect on the system itself. If you change all of the students in a classroom, it's still a classroom. While the students may look different, the system itself stays the same. Now, this isn't to say that these students won't perform better or worse, so elements are indeed important actors. But objectively, the classroom is still a classroom. It's the same thing as if you were to switch the frosting color on a cake. It looks different, but the system remains the same. We can thus conclude that systems can change elements and still remain recognizable.

You can change all the players on the Dallas Cowboys football team, but you will still know it's the Dallas Cowboys, just like how teachers change and get replaced at a school, but NYU still remains NYU. Even if substitutions to the elements of a system change, the system still remains itself. It might experience some minor changes over time, or it may not change at all. If the interconnections and purposes remain the same, the system will still perform the function it was made to perform.

Unlike elements, changes to the interconnections of a system have more visible impacts on the extent of unrecognizability. Let's say you kept the same students in a classroom. However, instead of learning, they are now in charge of teaching the teacher. Or if you kept the players on a soccer team, but changed the rules to those of handball—the system is going to look a lot different. It might be fun to watch students teach teachers and a soccer team throw balls to each other, but the system is going to be completely different. You wouldn't even be able to recognize it as the same system as before. A soccer team is no longer a soccer team if they touch the ball with their hands. So changing interconnections is a much more lt and impactful transition than changing the elements.

A change in the function or purpose of the system can cause the biggest impact. For example, you keep the students and the interconnections of a school, but its purpose would not be to learn, pass exams, and get a good job. Instead, it's to avoid learning, to fail, and to end up jobless. Or the function of a cake would not be eating it, but to tear it to pieces and throw them on a canvas to make art. If the elements and interconnections stay the same, but the purpose changes, the whole system can change. But let's take this thought to the extreme. Donella Meadows in her book *Thinking in Systems* gives the example of changing the function of a tree to take all the nutrients and grow gigantic instead of just survive and reproduce. That would change the (eco)system entirely.

To answer the question "Which part of a system is the most important?" the answer is: all of them. Elements, interconnections, and purposes are all needed in a system. They rely upon each other and they each have their job to make a system work.

If we asked the question "Which part has the greatest impact on system behavior if changed?" the answer would be purpose or function. Purpose or function, the most abstract part of a system, changes the system the most if it gets changed. Interconnections would be next, because a change in a relationship will also change the system behavior. The part with the least impact is the elements—strangely, the most visible part within a system. You can change the elements without changing the systems behavior.[x]

Why is important to know which part has the greatest impact if changed? For two reasons. First, if you experience disturbance in a system, you can now know on which level to seek the error that needs to be fixed.

Second, if you wish to change a malfunctioning system, now you know what impact your interference will have on different levels. For example, if you have a company which experiences a deficit for years, you know that just changing the employees won't make a big difference unless they will also have a better work ethic (change on the interconnection level). You can also change the function of your company—you can downsize the company to one worker, yourself. Thus, you can save on employee wages, taxes, etc. However, this way your affair hardly can be called a company, rather an individual business.

Other Expressions

It is easy to think of a system in relation to its elements, interconnections, and purpose or function, but there is far more to consider if you want to see the whole picture. We can say that the three parts we learned about so far are the big parts of a system. Now let's familiarize ourselves with the other expressions to deepen our understanding of systems thinking. These are the smaller parts of a system according to Donella Meadows.

Stock

The first part you should know about is the stock. Stocks are the foundation of a system. They are the system's elements that "you can see, feel, count, or measure at any given time."[xi] Imagine it as the icing on a cake, the people in a building, the cash in your wallet, or even the water in a lake.

It's important to note that stocks don't necessarily have to be physical. It could be a person's confidence, or a flow of information.

Flow

Flow is the action that causes stocks to change. It's the growth and death of a flower, the debits and credits to your bank account, the number of times you succeed and fail, or the rainwater and sun evaporation in the lake. Donella Meadows refers to the relationship between stocks and flows when she writes: "Stocks are the memories of the history of changing flows within the system."

When there are more inflows than outflows, stock levels are going to increase. But if there are more outflows than inflows, the stock level decreases. As long as the outflows and inflows are equal, the stock will remain the same. Stocks are the buffers in systems and allow inflows and outflows to be independent.

For example, your bank account is a stock. If the inflow (of money) increases, then the account grows. If the outflow (of money) is higher than the inflow, your fortune decreases.

The flowers in your grandmother's garden are a stock. If she plants more flowers than what dies, or more than what she uses for decoration for Sunday lunches, her stock will increase. This means the flower inflow is bigger than the outflow. If more flowers die or end up in a vase (outflow) than what she plants (inflow), her stock will decrease.

Understanding the dynamics of stocks and flows and how they react over time is the easiest way to figure out the behavior of more complex systems. Imagine you are making a cake and you have your flour ready with your empty bowl. Fill the bowl with the flour and then do not touch it at all. The amount of flour will remain the same. It's static.

Now, imagine that there is a hole in your pot and the flour slips out. Obviously, the flour is going to end up gone from the bowl and the bowl will be empty after a while. But if you fill the bowl up with more flour at the same rate as it's slipping through the hole, the bowl of flour remains at the same amount because the inflows and outflows are the same. This is called dynamic equilibrium because it is not changing, even though there is something going out and something coming in.[xii] If you start pouring the flour more aggressively into the bowl, the inflow will be higher than the outflow through the hole, and thus the bowl will slowly start to fill up with flour. If you stop pouring flour, the inflow, the level of the flour will start decreasing again.

Let me ask you a question: While you were reading this paragraph, what did your mind focus on? The flour and the bowl with a hole on the bottom, or the actual flow? If you are like most people, your imagination was more set on the objects than the event itself. In other words, you were focusing on the stocks, not the flows.

When we do focus on flows, we seem to focus on inflows more. For example, pouring the flour more quickly, or working more hours to get a higher income. We often fail to realize the bowl of flour or our income can raise in two ways. In one way, we increase the inflow, but we can achieve the same result by slowing down the outflow. Covering the hole on the bowl or spending less money can bring us the same results. Stocks can't change immediately. The bowl won't fill with flour within seconds just because we covered the hole. We won't accumulate $10,000 in savings overnight (unless you win or inherit something, but that's another story). It's important to understand that systems behave the way they do because stocks take time to change because flows take time to flow. It can take a long time for a stock to increase or decrease. The population of a country won't increase or decrease overnight. Neither will the stock of a forest. Trees and people take time to grow, right? The changes in stocks set the pace for the system's dynamics.[xiii] Sometimes you can use the momentum changes create to work toward a desired outcome.

People who understand systems thinking view the world as a number of stocks, measure their natural dynamics, and based on what improvement they want to achieve, they interfere with the flows to lower or raise the stocks. This is what leads us into the different feedback processes we already talked about in the previous chapter.

Feedback loops

Stocks don't increase, decrease, or stay static by choice. Flows do not happen by chance. There is a control mechanism behind the process. This control mechanism operates due to the phenomenon we call a feedback loop. The same behavior over time must be triggered by something. When the answer is yes to this statement, we can be almost sure that there is a feedback loop behind that repeated behavior.

"A feedback loop is formed when changes in a stock affect the flows into or out of the same stock," Donella Meadows writes.

Let's take the same example of your bank account—your savings account, more precisely. The bank gives you a certain amount of interest at the end of each year. The amount of this interest will depend on the amount of money you have in your account. The more money you have, the more interest you'll generate. The percentage of the interest is fixed, but the actual amount of money you'll receive depends on the "size of your stock."

As we established in the previous chapter, there are two main types of feedback loops—reinforcing and balancing. What's happening to your savings account is the result of a reinforcing feedback loop. This type of feedback loop is responsible for growth or decay. How does it work on your savings account? Each year, you'll get a certain amount of interest, which will make your account grow. *Ceteris paribus* (if everything else remains unchanged), the next year, the interest you'll receive will be higher than the year before. It will be calculated based on your total amount of savings—which now is the total amount of a year before plus the interest of the current year. Thus, there is a slow growth each year. The growth is reinforcing itself each year.

Let's say in 2016 you had $1000 in your savings account and you get 10% interest (how amazing would that be, right?). At the end of the year, you received your interest: $1000 + $100 = $1100. At the end of 2017, you will get the same 10% interest, but this time it will be $110, so your total savings in 2018 is $1210. At the end of this year, your interest will be $121, and so on. There is an exponential positive growth. Of course, this is a simplified example. We assumed nothing changes: there is no inflation, no outflow, no changes at the bank, no charges, etc.

The balancing feedback loop balances. It's that simple. It's like how your digestive tract uses some bacteria (which are good) to keep out other bacteria (which are bad). The most popular example in systems thinking used to explain a balancing feedback loop is the behavior of a thermostat. If you set it to keep the room at 70F, when the outside temperature drops—and thus the room's—the thermostat heats. When the outside temperature—and thus the room's—rises, the thermostat cools.

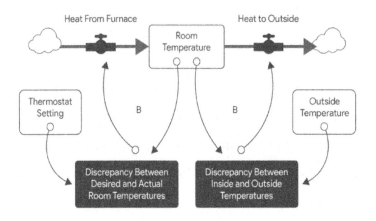

Picture 11:[xiv] *A one-stock system with two balancing feedback loops.*

As you can see in Picture 11, there are two balancing feedback loops (B) in action to regulate room temperature. As we discussed earlier, there are two ways to fix a problem—focusing on the inflows or the outflows. If we focus on the thermostat, we focus on the inflows. We can focus on the outflow by improving the insulation of our home so its inside temperature wouldn't be so affected by the outside temperature. Heat wouldn't vent out so easily.

Growth or decay happens from at least one reinforcing loop. Goal-seeking behavior happens from a balancing loop. When we observe growth in a certain domain which is followed by a decay (or vice versa), this indicates there was a shift in dominance between the two feedback loop types.[xv]

To understand the shifting of dominance in practice, let's take a look at another simple and popular model in systems thinking: population change.

Let's say a certain species of animal showed a rapid growth in a certain area. Usually, this is only concerning when there is an overabundance of predators, like coyotes or wolves. When there are too many of these predators, prey animals die off at an increased rate. However, environmentalists then have some population control on their hands. To stop the prey from becoming nonexistent, they have to do something with the predators. Relocating the predators, issuing more hunting licenses, or killing off predators then has an effect on the system that will increase the prey population because fewer end up on the proverbial plates of the predators, and more can repopulate. Let's use wolves as an example.

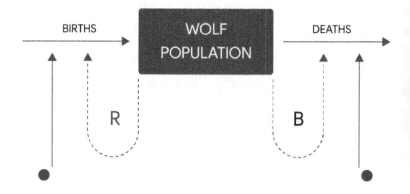

Picture 12: *The wolf population's dynamic change.*

Picture 12 is an example of reinforcing and balancing feedback loops. The increase of the wolves would be the reinforcing feedback loop (R)—the more fertile they are, the more members they will have to reproduce, and thus their population will increase. The balancing loop (B) would be the death or relocation of the predators. These two loops together represent shifting dominance. This is an important concept in systems thinking because when one feedback loop is dominant, there is an impact on how the system behaves. When fertility is higher than mortality, the amount of wolves will increase. We can interfere to stop them from growing by hunting them, relocating them, etc. If the interference is successful, the dominance will slowly shift toward the balancing feedback loop—mortality will increase, and if unchecked, will overpower fertility.

You should always double-check your data. There might be errors in your calculations or observations. Donella Meadows recommends asking the following three questions:

- Will the driving factors really unfold this way? (Are the birth or death rates likely to happen as I predicted?)

- If they do, is the system going to react in the way I think? (Will the increased amount of wolves really kill all the bunnies?)
- What is affecting the driving factors? (What encourages birth rate and what prevents death rate, or vice versa?)

The first question is a guess, a prediction about a likely future, so it can't be answered with facts. You can't prove that you are right, but asking yourself this can help you decide which future is most likely and needs to be seriously considered. When system analysts consider life-or-death predictions, they run a lot of tests and explore different scenarios and outcomes, as well as their consequences.

The second question intends to ask how good your model is. You'd have to dig deep and look into the model you use to double-check if it includes the correct system dynamics and so on. More detailed models may be needed to separate the population by age group, income level, belief systems, etc. With such a complex model, even if the numbers are not correct, the behavior itself can be predicted with high accuracy.

With the third question, you are essentially looking for system boundaries. You look at the driving factors and investigate whether they are independent or not. Is there something like economy, social beliefs, or financial status affecting birth rate? In the case of wolves, are there any environmental changes, climate changes, or human interventions that make fertility or mortality rate changes?

Fluctuating Loops

In this section, I would like to introduce a new aspect of system behavior that affects the activity of feedback loops. These are delays.

Delays cause the feedback loops to overshoot or undershoot a goal. What this means is that by the time someone realizes something needs to change, it's too late. Let's go back to the example of the thermostat. Even if you set your thermostat to a certain temperature, 70F as we said before, the house will almost never be that temperature. The heat decreases quicker in the room whenever the outside temperature drops, bringing the inside temperature of the home beneath the programmed temp. If the outside temperature drops more quickly than the thermostat could heat up the room, you won't be able to reach the desired 70F.

There are also information delays within a system. In the real world, even in our high-speed one where information flows quicker than ever before, there are delays. Think of the airline companies. They overbook their flights because there are often a few passengers who end up not checking in or taking the flight. It's their buffer, and it usually doesn't cause a problem. However, sometimes the flight is overbooked and it does cause a problem.

With this example, there are three delays. First, there is a perception delay. The airline isn't going to react because one flight truly was overbooked.

Second, there is what's called a response delay. Even if the airline sees a trend of flights being overbooked and causing an issue, they can't fix them all at once. They will start refunding or canceling on some passengers and moving them to another plane. If they find another flight to the desired destination with empty seats, they just try to move the affected passengers there. They make cautious adjustments just in case they made a miscalculation about the trends.

Third, there is a delivery delay. Even though the airline wanted to correct its mistakes, it takes time to refund, relocate, or compensate the passengers. They can change their policy on overbooking, but it still takes time to resolve the issue of the already overbooked flights.

Because of perception, response, and delivery delays, systems deal with what are called oscillations. Oscillations are like a rollercoaster of traffic or sales. In the airline example, when they stop overbooking their flights, soon in the future they will experience a growing trend where they have too many empty seats. What do they do? They start overbooking their flights again. There will be a period of optimized flights with no extra passengers, but soon they'll run into the issue of having too many passengers per flight. There is a precarious balance that needs to be reached, but it can be hard to understand the trends when these three types of delays are in action. The oscillations can be diminished with different strategies. However, in our case, it is more important to understand why they happen. Airline companies are not dumb or hungry for money necessarily. They only wish to optimize and maximize the profits of their company—just like any other sensible business does. They are actors in a system with information delays they can't influence.

Systems analysts are in tune with delays because changing delays can make the system easier or harder to deal with. Because of this, they are always on alert as to where the delays are occurring within a system. Unless you know where and how long the delays are, you can't begin to understand the behavior of the system.

When you look at the airline example above, you might think that it's easily fixable. But imagine how many flights every airline is dealing with on a daily basis. If every flight is overbooked or underbooked, you suddenly have a catastrophe, and it affects everything. If the airline is making less money by underbooking their flights, their employees get paid less, repairs don't get done, and they end up losing large amounts of money. On the other hand, if they overcompensate and overbooking flights becomes the norm, suddenly employees are dealing with very upset and angry passengers, the planes are being used more often with fuller loads, and the company's reviews go down.

Chapter 4 - Understanding Bottlenecks, Leverage, and Feedback Loops

Bottlenecks

The first model we'll talk about is called the Theory of Constraints. This theory talks about how every system is limited by different constraints. One constraint, however, is going to be tighter than all of the other ones. The constraint that is the tightest of all is called the bottleneck. It's "the point of greatest congestion that is causing a delay within a system."[xvi]

The theory of constraints implies that the system's performance is going to be limited by the bottleneck. If there is no change to the bottleneck, the system will not improve. When the bottleneck is influenced, the system changes.

We can think about the theory of constraints in the context of our own lives. Many of us don't know what our bottlenecks are. If we fail to address this key issue, we'll fail to move forward in life or see anything change. We don't need to work harder on our self-development; we just need to identify where to concentrate our efforts.

You struggle in your relationship. You lash out at your partner often without knowing why and you feel insecure, which exacerbates your number of your outbursts. You read books on anger management, practice meditation—you pay for a Tony Robbins seven-day seminar using all your savings. You do everything humanly possible to become a more patient and kind person, but after a month of chill and post-Tony Zen, you lash out again. Why? Because all the humming on the mountaintop, all the mumbo jumbo, and all the optimal personality growth were just symptoms of your main issue: your early relationship with your dad.

Once you can stand in front of him, look him in the eye and say, "I understand that I ended up with this mentality because you told me this is the right way. You were never really there for me except when dispensing a few encouraging words, but leaving me alone in the matter nevertheless... I learned through the years when I couldn't rely on you to believe that one can't rely on men. That they often lie, as you did. I know that you gave me the best you could according to your best knowledge. It will be a long journey to change, but I'm up to it. I forgive you."

The bottleneck won't disappear. Your dad won't change, probably, and your upbringing won't be erased, but now you'll be able to move on and truly change. Now you'll know that you're not a crazy person.

Every system error has a bottleneck—economic, political, and/or social issues. There is one problem which, if treated, can make it easier for the system to change and improve. It is not a guarantee that anything will change in the end. But it is a guarantee that as long as the bottleneck issue is not addressed, nothing will change.

Leverage

Leverage is a skill to influence a system to "return the maximum effect per a unit of effort." Today, our attention is scattered and our time and energy limited compared to how many options we have. However, we can maximize the benefits of our time and energy by using leverage.

What do I mean by that? If we consider that time, energy, and focus are interchangeable, we can "exchange" them to get the most benefits. For example, sometimes you can use more of your energy and focus to win time. Let's say you need to leave work two hours early to attend your kid's graduation. You can invest more energy and focus to finish your daily tasks at work to "win" enough time.

The Pareto Principle, or the 80/20 rule, talks about how 80% of one's results can be achieved with 20% of one's effort. Most of us already know that, but that's not what's truly important in the image you can see below. Instead, if you move down the line, you'll see that "50% of one's results can be achieved with just 1% of one's effort."[xvii]

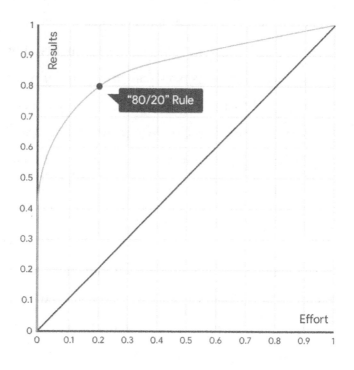

Picture 13:[xviii] *The Pareto Principle—differently.*

You should want to be on the straight line. Some tasks don't require all our expertise, and the marginal return any extra investment of time, energy, or focus brings is simply not worth it.

My friend and role model, Derek Sivers, told me a story once. He used to bike down Venice Beach every morning when he lived in LA. He was kind of competitive about it. He was biking as hard as he could, huffing and puffing, investing all his might and energy in the process. He went up and down the beach in 43 minutes. He didn't enjoy it. Every day, it was more and more burdensome for him to go out and bike because he knew the physical pain he'd feel. So one day, he changed his mind. He didn't rush; he casually biked, enjoying the view. When he was done, he surprisingly experienced that he'd ridden the same distance in 47 minutes. All that suffering and huffing and puffing was for an extra four minutes.

Many of us spend too much effort and don't get the desired results, or we feel like the outcome was not worth the effort, as Derek did. You have to learn what level of expertise is needed to hit your goals. Donella Meadows has three things that are considered to be the most important leverage in a system.

The first is changing the rules. This can help define what is possible within the system. What rules can you change in your system called life? Your habits, what you invest energy and time in, which area of your life you focus on… these can all optimize your efficiency.

The second is building in self-organization. This means designing your system in a way that it will improve naturally over time. These can be self-made constrains and checkpoints—for example, setting your alarm clock for two p.m. each day to ask yourself, "Am I productive?" It can also be deliberately not buying sugar or chocolate when you are on a diet, packing away your videogames into the attic, and so on.

Finally, the third important leverage is improving the information flow. We can reflect more accurately on our progress if we introduce objective and accurate measuring tools like feedback loops. They are loops because they are inspecting information as a circulation, not in a linear nature. The feedback itself is a tool of information delivery—it informs how a system is doing relative to the goal of the system. For example, if you are constantly arguing with your partner about nonsense, you might be in a reinforcing negative cycle. You need to be aware of this negative loop you are in to be able to talk about it and break out of it.

What you'll want to do is minimize delays between measurement and improvement. Imagine yourself as a product that needs improvement. It should be fast and consistent. A daily testing, measuring, and installing of improvements on your system can create a better leverage. If you work on a daily basis to improve your relationship, for example, it will normalize quicker.

So how do we use the knowledge we've gained about bottlenecks, leverage, and feedback loops? When we identify a bottleneck, we should devote our time to using the highest leverage until the bottleneck is no longer a problem. Then we measure our improvement-detecting feedback loops. Then we move on to clear our next bottleneck. The more bottlenecks we overcome, the less constraining the next bottleneck will be. This is a cycle where each repetition will push us to a higher level of emergence. "Shortly: find your bottleneck, experiment to remove it, repeat."[xix]

Chapter 5 - Problems That Make the Adoption of Systems Thinking Harder

What's Not a System?

We've talked about what systems are. There are systems everywhere you look. However, not everything can be classified as a system. So what's not a system? Anything that does not have interconnections or a purpose/function cannot be considered a system. Imagine the grass clippings left behind from a mowed lawn. They have no purpose. They just sit there or get taken away by the wind. Systems can change and adapt. They don't always have to be living things, but they are at least dynamic, goal-seeking, and adaptive over time.

Let me put it this way: People walking randomly through Times Square are not a system. People dancing in a flash mob in Times Square are a system. Why? Because they have some kind of relationship that connects them—the dance. Their purpose is to move together and have fun.

Problems in a System

When you think of a human purpose, it usually has a little bit of oomph behind it. Systems' purposes often are not like that. What's more, systems often produce purposes nobody really wants. They can also adopt purposes that only a few elements want, but their voice together grows and overtakes the system. For example, no one wants to create a society with high drug usage. What's more, societies constantly fight against drug use. However, it still *does* happen. How? Consider the following points:

- High depression and hopelessness rates in our modern society produce people who want quick relief from their pain;
- Farmers and dealers want to make money;
- Governmental restrictions criminalizing drugs and police who execute governmental orders both make drugs difficult to access, inflating drug prices and encouraging dealers to sell even more; and
- Non-users being more interested in protecting themselves than helping addicts recover

As you can see, solving drug problems is very complicated. Although the purpose of the system is to stop people from getting addicted to drugs, subsystem interests, like dealers wanting to maximize their profits or non-users wanting to protect themselves, sabotage the purpose of the system.

Some say that legalizing drugs is the solution. If drugs are easily accessible, they won't be so expensive, and thus dealers wouldn't be so keen to sell them, people wouldn't run into fake (and much more dangerous) drugs, and so on. The problem is that when we talk about drugs, we don't only talk about marijuana, but heavy drugs like hashish, cocaine, and other narcotics. Who would be insane enough to legalize those? Even with marijuana legalized, heavy drugs and dealers wouldn't disappear. They would only lose one of their products, so they would focus more on selling those drugs that they have left—the truly harmful ones.

What's the solution? If I knew, I would be holding speeches at the UN's General Assembly, not writing books.

This being said, there is an example of drug handling which, while it may not set a rule, it is certainly noteworthy. Portugal decriminalized all drugs in 2001, treating their usage a health issue and not a criminal one. The rate of Portuguese drug users and the rate of drug users dying from overdoses both dropped as a direct result. Portugal has the second lowest rate of death caused by drug overdose per million people in Europe.[xx]

Let's take a look at another example: low-performing schools. No one wants to produce or attend a low-performing school, yet they still exist. Why?

- Low socioeconomic status;
- The school is located near a neighborhood known for gang violence or drug abuse;
- Teachers lack higher credentials because they are not paid enough;
- The government aids are focused on schools that score well;

- The school doesn't have the funds to bring in better resources; or
- Wealthy people around the school either transfer their children or don't care to invest in the school

The U.S. has school zoning. This means that the school your child goes to is based upon your position within a certain zip code or codes. Schools located within zip codes with high property values and wealthier families consistently outperform schools which are zoned for zip codes containing lower-income families. And because lower-income families can't just up and move zip codes like the wealthy can, low-performing schools stay low-performing.

Unfortunately, these factors would be very hard to overcome, and the system could be trapped in an outcome that no one originally intended for or even wanted. The theoretical solution to system and subsystem antagonism is keeping their purposes in harmony.

The Things We Cannot See

You can look at a person and see their blonde hair, blue eyes, and height. But there are also things you cannot see, like their pain, struggle, or even what makes the person happy. Systems are a lot like that. There are problems that arise when systems thinking is not used in the right way. There are some parts of a system that we can't see. When we investigate the parts of a system, we often don't dig deep enough and only observe surface elements. Thus, we end up managing or improving only those parts of the system we can see. But as we learned in the previous chapter, if the bottleneck issue is not treated, there will be no significant changes.

Imagine a dog that just won't sit for the life of him. We can only manage the things that we can see with the dog. We give him treats, we incentivize him with strokes, kind words, and pep talks that only the craziest owners can do. Or we raise our voice, trying to intimidate him with "sit, sit, sit" and explode like Hades in Disney's *Hercules* when the dog tilts his head and wags his tail for the thousandth time.

We've tried everything to manage the problem we could see, but the dog doesn't sit. If we really looked, however, we might find that the dog hasn't had proper training, the reward system during training isn't very good, he hates our treats, he was never taught to sit as a puppy, or there's another dog in the house that also does not like to sit or listen to commands.

Sure, some of these might be the owner's excuses to explain his or her incompetence at teaching a dog to sit. But some of the reasons might be real. We can't know for sure. We have to experiment to find out what's real and what's not. The bottom line is that it's incredibly hard to manage something if you cannot or don't see the real issues.

Why Isn't Systems Thinking More Popular?

You're probably thinking that if systems thinking is as great as I'm presenting it, why isn't it more popular? Why aren't there more organizations, political entities, and regular people that are adopting it? It'd be wrong of us to chalk it up to a problem with systems thinking itself. Instead, it's really a problem with society's unwillingness to change.

Systems thinking is a relatively new discipline. In politics, for example (but also in economics and other large disciplines), the actors are thinking on a state level and focus on maximizing their power (or profits, in the context of economics). This is what the school of thought of realism is teaching us. *"Bellum omnium contra omnes,"* or "the war of all against all," as Thomas Hobbes, one of the early fathers of the realism school, says. But realism, while it was an efficient tool for predicting political outcomes, wasn't able to explain the behavior of the states. This is where systems thinking came into the picture—a complex synthesis of interconnections, behavior, and outcomes.

While the roots of systems theory date back to antiquity (Mayan numerals, Sumerian cuneiforms, Egyptian pyramids), the first comprehensive, disciplinary book on the subject, *Society for General Systems Research* by Ludwig von Bertalanffy, William Ross Ashby, and others, didn't appear until the 1950s. This means that it is a hardly-70-year-old discipline which still meets with criticism. Thus, changing to adopt a relatively new discipline, especially on the international stage, is risky.

There's also a mental health stigma against systems thinking. A lot of autistic people and people with Asperger's, in particular, are hardwired for systems thinking by default. It's one of the reasons allistic (non-autistic) people don't understand how they draw their conclusions, but why many of them excel in areas which require systems thinking (STEM fields, code breaking, etc.).[xxi]

One reason so many of us fear change is because we are told it's bad to make a mistake. All throughout our lives, there is this notion drilled into our minds that mistakes hold us back. August Bush III, a CEO of a highly successful company, told the rest of his vice presidents that "if you didn't make a serious mistake last year, you probably didn't do your job, because you didn't try anything new. There is nothing wrong in making a mistake, but if you ever make the same mistake twice, you probably won't be here the next year."[xxii]

Mistakes help us learn, and if we actually learn from them, people often forgive us. When you do something correctly, you're not learning anything new, because you already knew how to do it. It is not difficult to repeat or copy whatever you have mastered, and it is not going to change the world. As my father said, "No praising for basics."

But let's examine mistakes more closely. In systems thinking, there are two types of mistakes: commission errors and omission errors. Through these two types of errors, we can see why organizations (and people in general) aren't framing mistakes as opportunities to learn.

Commission errors happen when someone does something that shouldn't have been done, like when you eat that fifth cupcake and then feel really sick. Yep, that's when you know you've made a mistake. Obviously, this can also apply to big things like when companies make the wrong decision by hiring a certain employee, merging with a bankrupt company, or making a bad investment.

An omission error is when someone doesn't do something that should have been done. On the international stage, let's take Nokia's example. They were the leading mobile phone company in the 2000s. In Europe, at least, everybody had a Nokia. Nokia was the cool phone, the first phone to take a horribly pixelated, but hey, still a picture, and so on.

Then smartphones appeared. What did Nokia do? Nothing. When you could get an iPhone 3, Nokia was still fooling around with twelve-buttoned Methuselahs. They completely fell behind because they failed to react to the changing trends and tailor their products to meet the current demand. Today, Nokia lost its former dominance on the mobile phone market even though the company still operates and in the past two years faced an increase in demand. But even my dad has an iPhone.

Omission error is usually the more important mistake because it's most likely the cause of a failed or deteriorated organization.

Western end-of-the-year accounting summaries for organizations, just like we did in our self-assessment, track commission errors, which are not the bigger issues. Commission errors are usually more visible. For example, we assess our behavior as such: "I was stupid because I spoke ugly with my supervisor." We don't say, "I was stupid because I never took the time to learn to control my anger."

Everyone avoids making mistakes, and since businesses don't take into account the omission errors, they go unnoticed. Adopting a systems thinking perspective would force them to reevaluate both types of errors, and that's a lot of change for people who fear it. Someone in a manager position wants to minimize their mistakes. Since omission errors are often unacknowledged, all the manager has to do is minimize or transfer the blame of their commission errors to someone else. How can you do this? By doing nothing. Who doesn't work won't make any mistakes in their work, as they say. The less a manager does—and here I mean innovatively, as I mentioned before that repeating old tricks is not risky—the smaller their chances of making a commission error. This is a big reason why organizations don't often make drastic changes—nobody wants to take responsibility for it.

Another reason why managers of organizations or even regular people don't adopt systems thinking is because they simply don't know enough about it. There aren't many books or resources out there for the everyday person, and so people go without the knowledge that their organization or life could be improved. System thinkers usually produce material and discuss their profession for and between each other. While it makes sense (it's not the easiest topic to bring up at a dinner party), making systems thinking more popular could lead to better-running systems overall. Reading about systems thinking can sometimes feel like you opened up that old college textbook about nuclear fission. It's difficult. Then, when you do finally understand it, talking to someone else about it seems about as impossible as flying like a bird.

I get it, which is why systems thinking should be easier to understand. This is my mission with this book: to talk about the main systems thinking principles in an understandable language while keeping the gist scientific, making it easier to people adopt this mindset. Unfortunately, the discipline of systems thinking is filled with jargon. Many words sound fancy, so fancy that many of their users don't understand the meaning behind them, yet still use them with authority. For everyday people, systems thinking can seem like some abstract, high-class, academic-genius-requiring discipline which would take too much effort to grasp. "I didn't understand math in school, so there's no chance I will get this," the average reader may think. They may have a good sense of how to look at things from a systems point of view, but won't even try it when they hear stuff like "the emergence of reinforcing feedback loops in an interconnected environment."

Chapter 6 - How to Shift to Systems Thinking

Recently I watched a National Geographic series called *Genius*. It presented the life of Albert Einstein from his infancy to his death. Beyond the captivating story of his life, I observed even more interesting patterns. Einstein was a nonconformist, an original thinker who didn't back off or get intimidated just because people didn't agree with him once, or even ten times. He persisted with his work in physics, thought about things no one ever had before, worked relentlessly to prove his theories, and yes, he won a Nobel Prize for his work. Einstein loved thinking of quantum physics, but he didn't like it in its current form because it didn't make sense to him. Since Newtonian physics didn't answer his questions, he had to start and define his own answers. So he became a quantum thinker. He didn't disagree entirely with conventional physics; he just believed that quantum physics (his quantum physics) answered his questions better than Newtonian physics did.

He was a genius. Based on the series, he wasn't necessarily a kind, affectionate, or faithful man. But we have to separate the art from the artist, right?

Isaac Newton, father of Newtonian physics and one of the most influential thinkers in human history, was a bit of a nutcase too. He was a famously vindictive and petty man. When he worked, he entered in such an aggressive state of flow that he didn't eat or sleep for days, which of course made him unstable and mean. He often fell into depression, contemplated suicide, and had imaginary friends (or enemies, based on his mood). He also laid the foundation of classical mechanics, contributed to optics in a groundbreaking way, and yes, he is the father of the laws of motion and universal gravitation.

Another great genius, Nikola Tesla—who contributed to the world over 200 inventions including the first prototype of an electric motor, the first X-ray photography, and a more efficient electric system than Edison created—was a germaphobe. He had OCD, which compelled him to do everything in multiples of three, and he also suffered from hallucinations. (As an aside, Edison got so envious of Tesla's success that he relentlessly tried to destroy his career.)

Do you see a pattern in these stories? This is a little systems thinking test. What is the common feature of these great thinkers? Let's see.

For one, yes, you are right. They all were a little bit loony. Seneca was right when he said that "there is no great genius without a tincture of madness." This was my obvious message in presenting these great thinkers.

But beyond their madness, there was a creative genius that made them realize they wouldn't achieve anything extraordinary if they kept thinking linearly. Thinking that these people are mad because they are geniuses (or geniuses because they are mad) is an example of cause-effect thinking. "This happens because of that." This was not a way of thinking these geniuses ascribed to; they realized that conventional cause-effect (or linear) thinking wouldn't bring them answers. They didn't become systems thinkers (although I'm sure they didn't call their thinking method that) because it was trendy or distinguishing to think that way; they became systems thinkers because they realized that the world (or the universe) is a cluster of many complex relationships and interconnections. This reasoning gives a much broader explanation and perspective on their behavior. This is a more systematic reply than just saying they were too smart, so they became loony.

If we dig deeper into our systematic analysis, we may also discover that geniuses in general are not understood in their time. We idolize them today, but when they were alive, they were ridiculed and accused of, and sometimes even burnt or hanged for, witchcraft. It was not easy for them to deliberately choose to think differently. Not many thinkers and geniuses had the financial background to devote all their time to thinking. Einstein, for example, worked as a clerk for many years and hated it. He was secretly developing his theories under his work desk or at home at night. He didn't have an easy start in life.

I give you the homework to think of three more reasons why these geniuses all had a shtick. If you wish, you can send your answers to zoemckey@gmail.com. I'm truly curious. You can use systems thinking elements, for example. What kind of feedback loops were affecting the lives or the works of these thinkers? What delays did they experience?

Now that we've finished this mental workout about systems thinking, let's see how you can actually change your thinking patterns. A lot of us are linear thinkers, which may have worked for us so far in life, but systems thinking can help us make more sense of what's going on in our lives on a deeper level.

You don't have to overcomplicate your life where linear thinking serves its purpose. If your car runs out of gas, you have to refill it. No gas (cause), no car (effect). It's a simple linear equation and perfectly accurate. No need to overthink this one.

Symptoms vs. Problems

Linear thinking focuses on symptoms, whereas systems thinking focuses on the problem. When you solve a symptom, you don't solve a problem. When you get a cold and are stuck with a stuffy nose, headache, and sore throat, you can take some cold medicine to help those symptoms go away, but in four to six hours, you're going to have to take that cold medicine again to get rid of the symptoms. What you're doing is removing the symptoms, not the problem, which is the cold virus.

I just recovered from the longest illness of my life. In Mexico, I came down with a very aggressive throat ache overnight. What did I do? I went to the pharmacy, bought DayQuil and NyQuil, some Strepsils, and lemon-honey tea. For a few hours, I felt better, but then fell back again to misery. Linear thinking is a lot like masking cold symptoms with random medicine. Instead of looking deeper for the problem, like going to a doctor to help identify the real cause of the illness—a sinus infection, in my case—and take the appropriate, targeted medication, we simply try to take away the symptoms and hope that will fix things. Sometimes it might, but many times, it makes things worse.

There are a few signs that will show you you're experiencing a symptom rather than a problem. One way you can find out is if the problem seems too small compared to the time and energy it's taking. You know when your roommate or significant other starts complaining about things that seem like no big deal? They spend hours talking about how you forgot dish soap at the store, and it isn't until much, *much* later that you find out they are actually mad at you for putting less effort into the housework lately.

In general, people aren't going to have a problem with the brand of the milk you pick up, or the few extra minutes it takes for you to commute home from work. These little things might take a lot of your time and focus, but they are usually just showing that there is a bigger problem under the surface.

Another way to tell if you're dealing with a symptom is when the problem is not going away. For example, my good friend was very insecure about her relationship. She needed constant reinforcement from her boyfriend that, yes, he loves her, and no, he won't leave her. They had this discussion every week for a year. Each time, it seemed she finally believed it, but somehow the topic came up again the next week. It was only when they went to couple's therapy that they realized my friend was acting insecurely because there was no stable male figure in her life growing up. Her dad died early, and her mom changed guys like someone changes clothes. All her life, she'd felt men always leave. After approaching her issue from this angle and with some months of therapy, she is now much more secure, and she and her partner have a normal relationship. Her insecurity about her boyfriend's loyalty was just a symptom of a deeper issue. The bottleneck problem was the inexistence of reliable male figures in her life.

Sometimes, problems aren't being solved because they involve hidden omission errors. There are some issues that organizations, for example, just don't talk about. Maybe it's sexual harassment, or maybe it's that negative balance everyone sees in the company bank accounts. If someone mentions a problem and it gets laughed off, they are going to be hesitant to mention it again.

Often problems don't get solved because no one solved that problem before and people are afraid of making mistakes (commission errors), so they do nothing.

If the problem occurs on a weekly, monthly, or annual cycle, it probably runs deeper than you think. Maybe your significant other starts getting furious on December 23rd every year. It's probably not the fact that your turkey was a little dry, but more likely the fact that they don't want to deal with extended family coming to town.

In life, there are a lot of stressors and frustrations that come up. This can cause anxiety, and if that's present, problems often get swept under the rug and they pop up on rainy days. Added stress or anxiety could be a reason you're solving symptoms rather than problems.[xxiii]

What Stops Systems Thinking

It is hard to stay aware of linear thinking, because many of us do it on a daily basis. Thinking in a linear manner is a habit; it's our second nature. In the following paragraphs, I will present the 10 signs that can help you detect your linear thinking patterns, according to Jim Ollhoff and Michael Walcheski.[xxiv] If you become aware of the signs, it will be easier for you to notice when you're thinking in a linear way and change them.

1.) One thing that indicates linear thinking patterns is the "quick-fix" mentality. Whenever we see a problem, we want to fix it immediately. While it's not wrong to quickly take action, you should be fully aware of what the problem actually is. Acting without knowledge of the problem could just make things worse.

2.) A lot of times, we try to use the Band-Aid approach. We put a Band-Aid on the problem and hope that it gets fixed that way. In reality, this just ends up covering up the symptoms while the problem continues.

3.) This planning style might seem like a good idea, but it is "notoriously linear." It's the idea of making the budget at the end of the year. This idea applies to companies more than it does individuals. Being profitable is great, but it can't be achieved with arbitrary deadlines. Short-term fixes are going to be unsustainable in the long run. Thus, breaking the greatest sweat to fix short-term problems is very anti-systems thinking.

4.) Another behavior we are often guilty of is responding immediately. This isn't the quick fix I talked about before, but rather when you panic and react out of anxiety. Gut answers seldom bring the best results. Slow, deliberate thinking with a calm brain and considering different variables will lead to a better decision.

5.) Sometimes problems arise that we simply don't care about. We come at them with an apathetic approach, which doesn't help. If there's a problem, it needs solving. Instead of groaning and wondering why you have to deal with it, be open and use your imagination and creativity to solve the problem.

6.) Analysis paralysis. There's nothing wrong with getting a little bit more knowledge, but information isn't going to magically make the problem go away. If availability of information was the answer, everyone would be a millionaire with a perfect body. Taking action has the power to change things, based on the knowledge that you gather.

7.) Shallow thinking. Fast and information-poor thinking is everywhere. Information sources dumb complex data down to linear thinking. Why did this happen? Because of the terrorists! Because of Obama! Because of Trump! Chill, mate. These are only elements of the system. We saw earlier that they, by themselves, can't make a big difference. There are underlying, systematic problems that produce these issues. Great tragedies don't happen because of individuals; they happen because of the relationship between these individuals, because of the system boundaries they need to operate in, and because of the discrepancy between system purposes and subsystem purposes. Maybe this sounds infuriating to you, but we can't break down internationally relevant, complex events as being the singular fault of one individual—it doesn't matter which party they're in.

8.) "My needs come first!" Selfish thinking is a losing strategy because it won't help the system you're a part of. "My kid comes first, and she needs the best of everything!" "My job comes first!" Sure, and the kid will grow up as an oversheltered, spoiled brat who will constantly feel attacked and harmed by the world. No one will help a selfish, bulldozer type of coworker. The world is a big place, and it can become awfully lonely if you are fighting alone against everything and everyone.

9.) Meeting everyone else's needs and sweeping problems under the rug is also harmful, and you'll end up avoiding all conflict until it gets so bottled up that you explode. Suppressing and masking problems will never solve them, right?

10.) Last, but not least, authoritarian management can put an end to collective creative thinking. When you or someone else is saying, "It's my way or the highway," this is purely linear thinking. It doesn't allow any open discussion, lacks creativity, doesn't help you or others see the problem's complexity, and certainly doesn't solve it.

Did you recognize yourself doing any of these? Don't worry, I'm not here to judge. We all do them from time to time. These points are not saying that we are stupid, but that we are human. Since we are human, we can think, assess the consequences of our actions, and decide whether they are desirable or not. And if they are not desirable, our amazing human brain can debate alternatives. By doing this simple, few-steps exercise, we get much closer to a systems thinking–approved answer.

But how do we make a deliberate shift to systems thinking? Here are some tips. I must highlight that this shift will take time and require a lot of mental presence.

1.) Don't blame. Rather, think about what is influencing someone to act in a certain way.
2.) Don't assume you know The Answer. Think that you know one of the solutions to the problem. Look for proof that confirms your solution. Then look for evidence that disconfirms it.
3.) Don't focus on a certain thing. Rather, inspect the variables which influence or affect that thing.
4.) Don't focus on the negative behavior. Rather, try to detect the motivations behind it. Negative behavior is usually a symptom of something greater. What is the bottleneck problem behind it?
5.) Don't focus on the words people say. Rather, observe how they are saying them. Read between the lines. What other words are they hiding by saying the words they say?

6.) Don't focus on what individuals are doing. Instead, take a look at the system's dynamics. What forces are driving the individuals to act in a certain way?[xxv]

7.) Don't run away from problems. Rather, think about why you want to run away in the first place. What are you afraid of? What's the worst that can happen to you?

Start to adopt this way of looking at events and things slowly. Introduce it gradually to your work life and personal life.

Chapter 7 - Solving Everyday and Complex Problems with Systems Thinking

In the previous chapter, we learned that when solving a problem, we tend to look at the problem or the symptoms of the problem. What we actually need to focus on is the ecosystem that this problem exists in. Something like global warming can't just be thought of as the world getting warmer. The problem lives within everyone's homes, the cars we drive, and the pollution we are creating.

Systems thinking takes a problem and looks at it as a whole. It's a broader way to effectively problem-solve. Jay Forrester in his book *The Fifth*

Discipline defines systems thinking as follows: "Systems thinking is a discipline for seeing wholes. It is a framework for seeing interrelationships rather than things, for seeing patterns of change rather than static 'snapshots.'"

Instead of seeing the photo of what the problem is, we can look at the patterns and interactions which make up the problem itself. Like Albert Einstein said, "The problems cannot be solved using the same level of thinking that created them." Problems around us are often dependent on external factors, like with global warming. Instead of looking at global warming as rising temperatures, systems thinking allows us to see the interrelationships between human interference, laws of nature, chemistry and physics, and how they affect global warming.

The Approach of Systems Thinking

Everyone talks about analyzing problems, and the word "analyze" literally means "to break into parts." We talked about systems thinking using synthesis to study how elements interact with each other within a system and with other systems. Instead of breaking down a situation into smaller parts, systems thinking expands your vision to a big-picture view where the problem lies. While this is extremely effective for complex issues, it can also work well for solving everyday issues that you encounter. What can we use systems thinking for in our everyday lives?

- Any complex issue where multiple actors are involved to help them see the airplane view of the problem, not just their individual parts in it. For example, family quarrels, workplace disagreements, etc.

- Problems that don't seem to go away even though there were attempts to solve them.
- Problems with non-obvious solutions.

On a greater, non-individual scale, systems thinking can help:

- Natural environmental issues;
- Competitive environmental issues;
- Political issues;
- Complex social issues; and
- Economic issues

How Do Systems Work?

In a few paragraphs, I would like to recap what we've already learned about systems thinking before we continue our journey.

- Systems are composed of a lot of parts that are interconnected with each other.

- Changing one part of the system (interconnections or function/purpose) will affect the entire system.
- How a system is structured is going to determine its behavior. This means that the system is more dependent on the connection of its parts rather than the parts themselves. A toothbrush and toothpaste individually don't make clean teeth. It's the relationship of the toothbrush, toothpaste, and brushing that makes teeth clean, right?
- System relationships are dynamic. They change over time, and thus system behavior is hard to predict.
- There are three types of delays (perception, response, and delivery) that can happen within a system, which makes systems mapping more challenging.
- The major behavior dynamics are controlled through feedback loops. There

are two types of feedback loops: reinforcing and balancing.
- Complex problems can't be solved with reductionist thinking. Systems thinking can solve complex problems by mapping out systems dynamics, emphasizing the effect of feedback loops, and coming up with various solutions and possible outcomes as consequence.

How to Apply Systems Thinking to Solve Problems

I've been saying that systems thinking can help solve complex problems, but now I'll explain the steps you can take to apply systems thinking to the problem at hand.

a.) Present the system visually.

The first step is you have to familiarize yourself with the problem. You need to understand it. What's the easiest way you can achieve this? By deeply focusing on the big picture instead of looking at the parts making up the system. This often requires familiarizing yourself with the stakeholders of the situation you are trying to fix and get their story. Ask questions, note the answers, and try to put together the pieces of the conceptual puzzle.

If you are a visual type, you can use concept maps. These are complex graphical tools that will help present the system in a more in-depth way. Concept maps enable you to visually demonstrate the elements and interconnections of a system. An example is below.

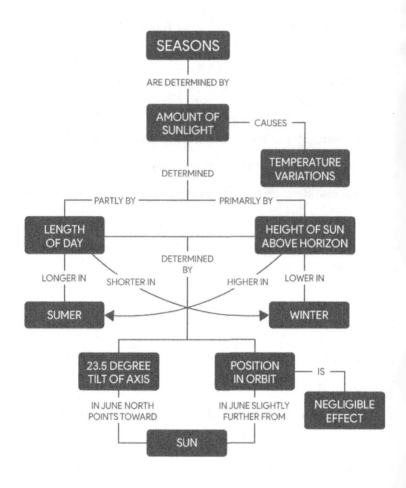

Picture 14:xxvi *Example of a concept map from Carnegie Million University.*

This concept map is discussing the relationship between the seasons and the sun. As you can see, there are a lot of natural variables in action which determine when and where winter or summer are. We can't really improve this system by human force, since this is a natural phenomenon as old as time. What we can improve, thanks to this information, is our preparation and readiness for changes in seasons. And I don't only mean buying the down jacket for winter; I mean, for example, getting ready for heavy snows on a country level by preparing the snowplows for action, and so on.

I went through a very funny situation in Hungary when I was a student. I was heading somewhere by train when a huge snowstorm rampaged across the country, making roads unusable. The Hungarian government sent out text messages to everybody—after the snowing began—to stay inside because of the hostile weather. The message came out too late and many people got trapped on the roads—it was a national holiday, so everybody was out to visit relatives. Many people were trapped in the snow for long hours—even days—because there were not enough snowplows and salt-spreading machines in service. The responsible authorities were very defensive about it, saying they were not prepared for the snow. Really? It is winter! What better time of the year to place snowplows in service than winter? Eventually Austria, our neighbor country, lent us some snowplows and cleared off our snow in half a day. Danke for that!

What could the responsible Hungarian authorities have done differently? They were not lying; they were not prepared for winter. Why? Probably because of some internal system error. People gladly pointed fingers at the Ministry of Environmental Protection and its leader, however, it was not only his fault. There were tax cuts and fund relocations that year, and thus certain institutions got less money. To not be forced to fire employees or cut their income, the institution had to save money somewhere. The maintenance of a snowplow is quite pricey, but since the past winters had been very mild, the ministry thought it would be a good idea to save money on those.

By the time the institution responsible for weather forecasting informed the Ministry of Environmental Protection of the storm, their funds were already depleted. They might have asked for emergency funding from the government, but they didn't get any. "We'll deal with the issue when its here." is a very common reaction to possible threats. When the problem happened, what did the government do? Spent millions of forints on text messages instead of resurrecting the snowplows.

We can see the repeated mistakes that were made in this situation. Who is at fault? No one and everyone. What are the influencers? Lack of money, lack of organizing, and priorities. There were a lot of linear thinking patterns involved in the case: we only do this if that happens, we pay our workers today even if the institution can't sustain its purpose for a nation-wide benefit, etc.

b.) Create a behavior-over-time graph.

After we have the system mapped out, the second step is to draw a Behavior-Over-Time (BOT) graph. Graphing the problem and how it changes over time helps us understand whether the applied solution is effective or not. An example of a BOT graph is below.

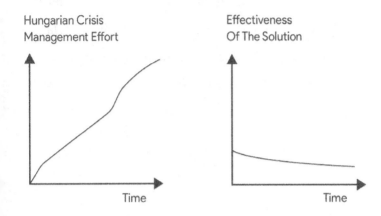

Picture 15: *Behavior over time in the case of the unexpected snowing.*

As we can see, as time passed, Hungarian authorities tried to take control over the situation. However, the cost and effort over time became less and less effective. They ended up spending a lot more money and finishing the job much slower than they would have if they'd maintained the snowplows in the first place. The military, volunteers, and the Austrians were all out to try to get a grip on the situation, but the cost-benefit ratio was not great. There was a big delivery delay involved as well. The government had to sit down and decide what to do, who to ask, wait for the Austrians to clear off their own snow, etc.

c.) Create a clear statement.

When you have a clear vision on what's going on in the situation you're looking at, what the main driving forces are, why the applied solutions are ineffective, and so on, it is time to come up with a statement that focuses on the bottleneck problem of why the situation is happening. This statement has to have a clear reason why the problem happens. In our case with the snowplows, the problem escalated to this level because of the lack of prioritized governmental funding and the lack of prioritized ministerial distribution of the funding.

d.) Identify the structure.

Now we have to identify the system's structure. This structure should include behavior patterns. These patterns will put the problem in a different light. Understanding the behavior and motivations of the governmental entities will help us understand how the problem could escalate to this level. Understanding the problem more can help us map out the system better.

e.) Get mile-deep in the issue.

It is time to get deeper into the problem. You'll need to clarify four things:

- What you want from the system;
- The mental models of the system;
- The broad view of the system; and
- What your personal role in the problem is.

The issue of the Hungarian environmental mismanagement is not the best to describe individual issues, especially the fourth point—your personal implication in the matter does not apply unless you have a decisive role in the Hungarian legislation corp. What you can do as an individual is volunteer to help save the people who were trapped in the snow. On a broader scale, you can choose to vote for a different government in case you voted for the current one.

For the example's sake, let's pretend you have an active say in this matter. In this case, we can answer the first three questions as well. What do you want from the system? To produce a better distribution and usage of funds (of which a high percentage are the contributions of taxpayers) to protect the taxpayers more effectively in the case of a national crisis.

Mental models of the system: What kind of systems archetypes are at action in this case, and how can the situation be mapped out the best way using systems thinking tools?

The broad view of the system: Taking a look at every actor involved in the problem. The government doesn't have sufficient money to distribute because tax evasion is high, and tax payment is low because taxes in Hungary are ridiculously high. The badly distributed funds are just a symptom of the main problem—that Hungary is not a rich country and doesn't have sufficient money to function independently.

f.) You come up with a plan of intervention.

Finally, you'll come up with an intervention based on all the information you collected and map out the issue. After clarifying everything and getting a better understanding of the problem, a solution should come that will aim to be effective in solving it for good.

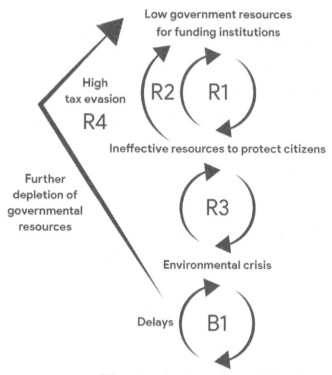

Picture 16: *Systems mapping on the issue of the Hungarian government and the snowplows.*

Picture 16 maps out the situation we were talking about in this chapter. We start with the standard problem presented by feedback loops R1 and R3. The Hungarian government doesn't give enough money to the institutions, thus the Ministry of Environmental Protection can't prepare properly for the winter, which we know is always coming, and when it comes, the ineffective resources make the problem even worse. Partially because the citizens experience that the government doesn't keep them safe despite the high taxes they pay, they start avoiding paying in taxes, which creates reinforcing feedback loop R2, which is partially the reason why the government doesn't have enough money to distribute (this is a simplified model, which doesn't take into consideration corruption, inflation, other governmental resources, international aids, etc.). Furthermore, the unexpected environmental crisis also depletes the government's budget because now they have to pay more money to solve the problem (feedback loop R4), as we saw in Picture XV about behavior over time.

There is, however, some short-term effort made to fix the current crisis (feedback loop B1) which is arriving with delays. The intervention to bandage problems in the short term creates a balancing feedback loop.

What's the solution? It is difficult to tell. If tax evasion decreased and taxpayers paid more taxes, the government would have more money to take care of its citizens. But citizens don't pay taxes because they feel they are not taken care of in exchange for the high taxes in Hungary. The issue is not so simple in this case, as the model shows. Citizens harbor some distrust toward the government due to corruption. Taxes and international aids get invested in low-priority products such as soccer stadiums.

Through the example of the snowplows, we examined one very specific glimpse of the Hungarian government's issues. Since this entire event is a minor problem within a larger problem, we can't deduce a comprehensive solution just from this one event. However, thanks to thinking the issue through systematically, we can see that there is more to the matter than just the Minister of Environmental Protection or lazy snowplow drivers. If we dug deeper and looked at the entire Hungarian legislation and execution system as a whole, we could find comprehensive ways to improve the situation of fund distribution and fund utilization. Just like in the example of drug problems earlier in the book, here too the system has the goal of serving and keeping its citizens safe. However, subsystem goals and interests just won't let the main goal of the system happen.

Another Practical Example of Systems Thinking at Work

Daniel Aronson, a systems thinking expert, in his work *Overview of Systems Thinking*,[xxvii] presented an example of using systems thinking through an action that was taken to reduce crop damage by insects. Just like we get prescribed antibiotics to combat a bacterial infection, the crops were sprayed with pesticides. The goal was to kill the insects. Now let's not take into consideration the soil, air, and water damage these pesticides cause. Let's just imagine that this is a magical pesticide that kills off all the insects when applied. So the farmers' logic would be this: "The more pesticide I apply, the fewer insects my crops will have, and thus the more crops will stay healthy and can later be sold."

Let's not be so quick to celebrate. While pesticides helped at first, eventually they made things worse. How? While in the short term the pesticide indeed was efficient, in the long term, there were side effects that no one thought of. The next year, the crop damage got worse and the pesticide seemed to lose efficiency. The insects that were feeding on the crops kept other insect populations in balance, either by competing with them or by preying on them. When the first crop-eating insect got out of the picture, the insects they controlled had no more competition or predators, and thus they were able to grow in population much more quickly and in greater numbers than before. There were now many types of insects damaging the crops. Since insect type A had been eliminated, farmers had to deal with a much larger number of insect type B. In the short term, Old MacDonald could keep his crops clean on his farm. But in the long term, he got into a worse situation than the year before.

What to do now? Applying the models of systems thinking, experts developed the Integrated Pest Management approach, a controlling group which is responsible for helping farmers with insect problems. However, they don't work with pesticides. They introduce the natural predators of the harmful insect in higher numbers to the affected area. The National Academy of Sciences tested and proved that this method has been more efficient in the long term, and more environmentally friendly too.

The two step-by-step complex thinking examples we talked about hopefully have shed some light on how you can explore the complex problems in your life or your organization. Looking at individual parts won't help solve complex problems.

With advancements in technology and human resources, problems today are unlike anything we have seen before. The problems we are dealing with are more complex, and they affect our lives tremendously. The complex problems are complex because they involve multiple entities and can be the result of our past actions. Solutions presented to us by others with traditional thinking methods will often be too simple to actually solve the problem. We don't want a Band-Aid fix; we want the actual fix. Whether you're dealing with relationship issues or environmental issues, systems thinking could be a better way of solving all your problems.

Chapter 8 - Social Problems and Systems Thinking

We are mostly aware of the big problems that are affecting our society; social, economic, and environmental challenges. With overarching issues like homelessness, poverty, and global warming, it's no surprise that everyone has an opinion on them. As Einstein said, new ways of thinking are required to solve old problems like these.

Social change is notoriously difficult to deal with. Even when we successfully implement interventions that work in the short term, they could create more problems in the long run. When we look at the problem with systems thinking, we can more easily see how the interactions of certain elements might sustain the problem in the long term.

David Peter Stroh talks about the application of systems thinking for social improvement in his book *Systems Thinking for Social Change*.[xxviii]

We do this quite frequently without even thinking about it. We solve a problem in the short term, but end up creating a more complicated problem in the long run, just like we saw in the case of crop-eating bugs and pesticides. Or, for example, if you borrow money to pay interest on another debt, you might solve your problem at first, but now you are going to have a larger debt with multiple interest payments. In the long run, it's going to make it more difficult to pay off your original debt.

David Peter Stroh emphasizes the idea that very often we are the reason for perpetuating the system we wish to change—unintentionally, of course. If we take a look at our actions through systems thinking, we'll see how our short-term, seemingly positive interventions create a negative long-term consequence. We talked about this effect before. It is the classic case when we keep on treating the symptoms instead of addressing the bottleneck problem.

It's a little like feeding a wild animal. Have you ever watched an animal show where there is an animal starving and dying in front of the cameramen, and you think, "How can that person just sit there and watch them die?" I remember watching a video of a polar bear that couldn't find food due to global warming, and it was dying from starvation. Many of the comments asked why the cameramen didn't give the polar bear some of the food from their own packs. The answer was simple: If the cameramen fed the polar bear, it might keep the polar bear from dying immediately, but it wouldn't solve the problem that the polar bear's food source has disappeared. Instead, the problem is due to global warming. Feeding a wild animal might take away the pain for that animal for a few moments, but in the long run, it creates more problems, like a dependency on humans for food.

The Question of Homelessness

Today, one of the largest domestic social problem is homelessness. Some people give the same argument when they are questioned about why they didn't help a homeless person: Giving them fish will keep them fed for a day. And then what? Unless they learn how to fish, they won't be helped out in the long term. Moreover, they will be enabled to keep up their homeless lifestyle without seeking a solution.

I don't necessarily agree with the people who say this. I feel my obligation to help my fellow human, even though I know I won't give them a long-term solution. The issue of homelessness is again a complex, system-level problem that can't be fixed by occasional, individual efforts.

Stroh says that we need to look at the discrepancy between what we hope a system will achieve and the results it is currently achieving. By looking at this discrepancy, we can use it as a powerful force for helpful change.

Previously in this book, I mentioned that the theoretical solution to achieving a systems goal is to synchronize the goals of the elements. Let's see what's happening in the case of homelessness:

- The main goal of politicians and elected officials is to keep the support of the voters. For this, they are containing the costs.

- Business owners want to keep homeless people away from their stores so they won't scare off customers.
- Homeless shelter providers worry about filling their beds to keep their funding.

The only way to sync the different interests of the actors is to find a common ground. This is easier said than done. There are two levels that requires alignment.

- The first level is aligning people with different interests with each other.
- The second level is to align the individual actors' highest goals with their own immediate, short-term interests.

If you've read any book on human behavior, you know that we are creatures who prefer instant gratification versus long-term gratification. Thus, no matter how noble our long-term goal may be, if achieving it would mean that we lose something in the short term, we'd need an extra effort to go for it, regardless. Adding to our self-sacrifice the fact that we have to cooperate with people who have totally different interests than we do often seems too much to ask.

It is very important at this point to help the actors make an informed decision to commit to the higher purpose—letting them know what it will take to get there. People have to know what they are signing up for to ensure their cooperation will be strong and durable. The actors involved have to reconcile their future noble aspirations with their current everyday reality—and how the latter will change in the short term to achieve the former in the long term. What are the short-term costs and benefits, and are they willing to commit to endure them?

Stroh talks about taking four steps to create this alignment:

1. Understand there are payoffs to the existing system. This is the case of the status quo.
2. Compare and contrast the case of the status quo with the case for change.
3. Create solutions that help both long- and short-term interests. Also make it clear that there might be a tradeoff involved in the change process. Meaningful changes sometimes require people to let go of something.
4. Committing to the idea of change by weakening the idea of the status quo.

1. Understand the payoffs to the existing system.

If we look at a system well enough, we'll realize that it works exactly as it's designed to work in the current moment. First, this sounds outrageous. Who would design a system that maintains homelessness? The answer resides in the behavior of the actors. That they are saying they want to do or achieve something, but they are acting differently. In other words, the current status quo is providing them with more visible and quick benefits, and thus they are enforcing that instead of long-term change.

Change is scary, and it comes with costs, so it's no wonder many people enforce the status quo. The current system's payoffs are quick fixes, like temporary shelters to reduce how many homeless are seen. But reduced visibility doesn't really solve the problem, does it? It is a quasi-solution—

shelter providers feel good about themselves because they are helping those in need and getting their funds. Funders feel good about themselves because they are helping—for a relatively low cost.

The leap people avoid is that the real cost of change would be higher financial engagement, learning new skills, offering jobs for homeless people, and patiently waiting for the return on the investment. In the case of homelessness, high financial engagement would mean building safe, livable, and affordable houses, instead of maintaining the shelter system, as well as changing the purpose and tasks of the shelter workers. Not many people want to have a former homeless person as a neighbor—the attitude of the citizens should also be changed. Also, the homeless people should be assured that they won't be ostracized and that the solution is permanent; they won't end up on the streets again—unless they want to. The system should also include certain restrictions and checks and balances. If word of mouth spreads that homeless people get new, well-equipped homes just because they are homeless, many people will feel tempted to become homeless to benefit from the governmental facilities.[xxix]

2. Comparing the case of change and status quo.

When you're coming up with an idea for change, it should include the benefits and costs of changing versus not changing. It's usually easier for people to comprehend this rather than the benefits of not changing and the costs of a change.

Asking people what benefits they perceive from the change can help to get them involved and in on the process of change. It is somewhat like copywriting strategies. First, paint how the ideal, desired future would look. Then contrast it with a certain, undesired future if no action is taken to realize the desired future. Then talk about the steps that will help the desired future come true.

The rosy future expectations:

- Reduced cost of social services and emergency teams like shelters, hospitals, and substance abuse centers.
- Reduction of unemployment cost.
- A good feeling of helping people get off the streets for good.
- The ability to get funds for meeting best practice requirements to ending homelessness.

The picture of the undesired future:

- More homeless people on the streets, lower quality of street life.

- Exponentially increased costs for social services, emergency treatments, and unemployment.
- Bitter feeling of not helping those in need.
- Shallow federal funds.

The costs of change are:

- Investment in permanent housing.
- Closing shelters and changing the function of shelter workers.
- Helping citizens overcome their fears having ex-homeless people in the neighborhood.
- Helping homeless people overcome their fears of falling back to homelessness again.

Let's take a look at the case for the status quo, just to have a clearer picture of what the alternative to change is:

The benefits of not changing are:

- Visibly less homeless on the streets thanks to the shelters, and thus better street life.
- A good feeling about helping the short-term problems of those in need.
- Shelters get funding from the federal government.

3. Creating solutions and making a tradeoff.

People wish to keep both the benefits of the status quo and the benefits of change. Of course, this would be the best-case scenario, and sometimes it's possible. As long as the overall system aims to end homelessness and encourages providing permanent housing quickly, it is okay to use short-term solutions while the long-term goal is reached.

Unfortunately, not all situations can be in the "and" category and keep all the benefits.

Oftentimes, there has to be a tradeoff. It'd be nice to not let anything go, but sometimes you have to get rid of something to gain something better. And sometimes, systems have to get worse before getting better. Failure to let the status quo benefits go can be one of the largest obstacles to implementing a change.

Lyndia Downie, a well-respected president of one of the best homeless shelters in the country, Boston's Pine Street Inn, convinced the shelter's board to change their mission from being an emergency shelter to a real estate developer and landlord. That's obviously a huge change, and it was one that not everyone was on board with at first. She said, "What might we have to give up as an organization in order for the whole to succeed?" They had to let go of their self-image of being a homeless shelter and replace it with something completely new.[xxx]

4. Make a clear choice.

Letting go isn't easy, but you can help those doing it if you weaken the case for the status quo and strengthen the case for change. This involves two steps.

The first step is to get people to stop and listen to what automatically appeals to them. Otto Scharmer, in his book *Theory U: Leading from the Future as It Emerges*, calls it "presencing." This means that you depict a desirable future and bring it to the present. This optimally makes people eager to make the desired future reality because they can connect with its importance better in the present.

The second step is to deepen the person's connection with the case for change. You want to help people envision the desired future.

- They need to separate what they want from what they think is possible.
- Individuals also need to focus on what they want and the results, rather than the negatives they don't want.
- Ask them what they are doing or feeling differently—as an individual and/or an organization.

These questions help to bring out what people want from the case for change, or their aspirations.

What to Do When People Aren't Aligned

It should come as no surprise that there can be situations where the four steps I mentioned don't do it for the group. While these steps usually help align even the most difficult individual, there is no guarantee. Sometimes, you won't find common ground, and if you don't, then try these steps:

- Collaborate indirectly by addressing the concerns of those who are not aligned to the case. Try to engage them in the process through third parties they are willing to listen to.
- Try to work around the unaligned individuals.
- Work against the unaligned individuals through advocacy, legislative policies, and nonviolent resistance.[xxxi]

One thing to realize is that it's not always important for everyone to agree at the same time. Only a small amount of people in agreement is often enough to create a momentum for a movement that will spill over later.

And of course, there is the option where people deliberately choose to keep the status quo versus changing it. This is a completely valid and okay choice as well. However, then these people should not complain later about why they didn't get the benefits of the future they have chosen to give up. Voting for the status quo means accepting its costs, not only its benefits.

Exercise: What about the Criminals?

Stroh gives the example of the long-term dangers of pressured criminal incarceration. There is a quota stating how many criminals have to be convicted. Even though sending more people to jail will keep the streets clear of criminals for a time, these criminals will be released eventually (system delay). In the meantime, they get totally alienated from how to live in a society, they are not aided properly to reintegrate, their mind is set to prison rules, and very often, they commit another crime. This puts additional pressure on authorities to seek incarceration and labels convicted people as evil and dangerous by nature.

The whole system gets reinforced in a negative way. How can we solve this issue?

Stroh recommends preventing or decreasing the possibility of people turning to crime in the first place by building strong communities that can provide a safe and loving environment. Thus, people won't feel alone and in despair, which lowers their inclination to engage in criminal behavior. Convicted criminals should go through a better, more profound, and caring rehabilitation to reintegrate them into society. They should be offered a standard job and provided with solutions to not turn to crime again.

Based on the analysis we did in the case of homelessness, create your own analysis and try to come up with solutions to solve the issue of convicted people. It is a brainstorming exercise to practice systems thinking in social scenarios. As a reminder, here are the four steps of the analysis:

1. Understand there are payoffs to the existing system. This is the case of the status quo.
2. Compare and contrast the case of the status quo with the case for change.
3. Create solutions that help both long- and short-term interests. Also make it clear that there might be a tradeoff involved the change process. Meaningful changes sometimes require people to let go of something.
4. Committing to the idea of change by weakening the idea of the status quo.

Good luck!

Chapter 9 - The Story of the Bins

A system you probably don't think about very often is the system of garbage collection. Yes, garbage. What happens when you throw every Styrofoam container and plastic bag into the garbage? While it might not be as big of an issue for you living in a home or an apartment, it can be a pretty big deal for larger corporations. One corporation, Sabre Holdings, together with Presidio Graduate School found out that their trash just wasn't what it should have been. They had a 36% waste diversion. This is a corporation that took pride in their sustainability initiatives, and they did great with energy and water. However, their trash was lagging.

The team got together to see how they could reduce their waste and create a better environment for their employees and the Earth. After talking, they decided to come up with a 60% waste diversion goal. However, the CEO wasn't happy with that. He proposed an 80% waste diversion goal, more than double what they were currently doing. This was going to require some big ideas, so they had to come at the problem from a different angle. They decided upon systems thinking because regular thinking methods wouldn't have been able to double their waste diversion efforts.

They launched what was called the Less-to-Landfill initiative, which kicked off on April 22, 2010. Their main focus was on keeping waste away from landfills and pushing toward recycling and composting. They claimed that the success of their initiative was due to the ongoing analysis of the waste system and its feedback loops. Their success was also attainable because their corporate culture truly valued sustainability and innovation.

Sabre Headquarters' Waste System

To do an effective job, they needed to find the high-impact leverage points that could quickly alter the waste problem. They defined four subsystems:

- Campus materials entry;
- Waste sorting and disposal;
- Compost processing; and
- Waste hauling.[xxxii]

Once they identified the system with its dominant subsystems, they needed to map out the feedback loops that were affecting this system. In our case, feedback loops mean the information flow on the interactions between the subsystems and their elements. Sabre identified four feedback loops in their system (Picture XVII).

The first loop, called "materials loop," is part of the campus materials entry subsystem the researchers identified. The first loop is also the interconnection "between the flow of materials entering the Sabre campus and the stock of on-campus waste." This is a reinforcing feedback loop, which would contribute to waste increase over time if no changes were made. The alteration of this feedback loop would result in overall waste reduction, so the researchers acted on it first. Interfering with this feedback loop reduced landfill-bound waste. Then they made further changes, decreasing the amount of materials entering the campus.

The second loop, called "infrastructure loop," is a part of the second subsystem we talked about, namely the "waste sorting and disposal" one. The feedback loop is the interconnection between the flow of waste and the waste disposal of the workers and the amount of trash cans. This is again a reinforcing feedback loop. If there were no increase in the number of trash cans and workers' waste-disposing awareness, the amount of improperly disposed waste would increase.

The third loop is the "waste diversion loop," and it is also a part of the "waste sorting and disposal" subsystem. It is the interconnection between the flow of "aggregate waste and the stock waste," both selectively and unselectively disposed ones. The improperly selected trash becomes landfill-bound. This is a balancing feedback loop with the function of stabilizing the selectively disposed waste before it makes it to the "waste and recycling hauling" subsystem or the "compost processing" subsystem.

The fourth loop is called "composting loop," and as you guessed, it is a part of the compost processing subsystem. It is also the interconnection between the flow of compost processing and stick of compost. This one is a balancing loop, which aims to process all compost-bound trash within this system.

Picture 17: Waste system feedback loops xxxviii

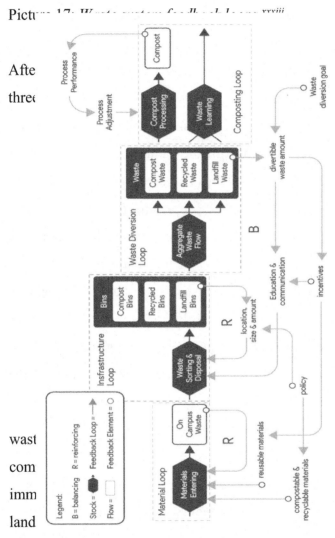

After
three

wast
com
imm
land
could actually go to compost or recycling. They

realized there was a lack of education among employees and decided to improve on that. They also included more signs by the waste bins and had announcements throughout the campus.

Unfortunately, two months later, their waste diversion was still the same. They went back to their feedback loops and analyzed what the issues at hand were. Most people are so used to throwing away items that could be put into the compost or recycling that many of the issues were mistakes. To make things easier, they had to implement more items to the workstations that would include smaller compost and recycle bins instead of the waste bin at every workstation. While this initially cost them $10,950 to add, they would end up saving $20,000 annually.[xxxiv]

Still, about 70% of the waste in the landfill bins could have been composted or recycled. That's when the decision was made, after checking the feedback loops, to standardize the colors of the bins, put up a goal thermometer, and add more signs and announcements throughout communications. Finally, they also had an awareness class that, if attended, the employees would earn credits for their campus health programs.

Without taking a systems thinking approach, this initiative would have never worked. It was extremely complex, and most people thought it was nearly impossible to more than double the amount of waste diversion. There were many times that Sabre had to rely on the feedback loops. For example, one feedback loop showed them that they weren't adequately prepared to undertake this initiative. They realized that the compostable cups were jamming the pulper, and the paper towels were too dry and took away from the natural composting process. There were simply too many paper towels, and the feedback loop helped them see that they needed to change the flow of the paper towels so that so many were not being wasted. This allowed the paper towels to be sent through the pulper with extra water and solved the problem.

The monthly waste diversion rate reached 75% by December of that year. A year from their kickoff date, they met their goal and exceeded it with 81% waste diversion. However, their rate dropped from 83% to 77% in June because of the new employees hired on. Their feedback loops showed that these employees were not trained in the waste management procedures like the rest, so new employees were offered campus tours, which greatly helped.

At the end of 2011, Sabre Headquarters sent 44 fewer tons of waste to the landfill compared to 2010, and increased recycling and composting by 122 tons. It just goes to show how much one corporation can do when they implement a change.

If you went into a business right now and asked them to more than double the amount of waste diversion they were doing, their eyes would probably bug out of their head. Okay, not literally, but this was a huge challenge to take on. Not only was it a great example of sustainability, but it was also an incredible feat in using systems thinking. This would not have been possible without the constant feedback they were getting from their feedback loops. Also, they engaged the whole system from the CEO to the lowest employees. By combining their top-down initiative with the senior management and employee engagement, they succeeded in their goal.

That's not to say that their system didn't have many challenges. It did. That's what is so great about systems thinking. Using this method helped them realize where and when they went wrong. Instead of seeing the problem as "Well, we aren't meeting the goal," they looked at the problem and asked *why* they were not meeting their goal. Having the system mapped with the feedback loops showed them when something was going wrong. They weighed all the benefits and costs of implementing this change, and they made it as a team.

The senior management staff ran into problems along the way and worried whether the employees would be aligned with the change. They realized they were asking a lot by making them essentially relearn how to throw something away. With hard work happening, the last thing this company wanted to do was put more stress on its employees when they were throwing away their granola bar wrapper or packed lunch. However, they also knew what their company's values were.

Sustainability was not something completely new to this company, so while the waste diversion increased greatly, it wasn't like the employees were unaware of what the company believed in. When they finished up and met their goal with systems thinking, Sabre Holdings walked away with five key insights.

Their first was that people come first. Change cannot happen without the help of the people within the business system. Focusing on the people helped them achieve success with their trash initiative. The CEO was the person who recommended the 80% goal, but he wasn't the only heart and soul of the initiative. The eco team leaders came up with fun games that would make the challenges fun and exciting to participate in while also collecting feedback.

The second key insight was that they really worked with the culture. The culture of the company and initiative became "we." Even their housekeeping teams and facilities management teams wanted this initiative to succeed, so they all worked together and were committed to the goal and achieving success.

The third key insight was how well the feedback loops worked to their advantage. It would have been impossible to make the initiative happen without the constant feedback they were getting. It helped them gain new insights, create new ideas, and communicate where they were in achieving their goal. The feedback helped to stimulate the rest of the employees to want to do more to meet the goal.

The fourth key insight was that systems thinking helped them look above and beyond what a normal way of thinking would have. They called it "more than 20/20 vision." This holistic approach to change via systems thinking helped them look beyond the immediate department of facilities management. They knew it involved everyone around them and everyone who would interact with the system. It didn't mean they had to change everything, but they did have to understand how other systems were affecting their own. They encouraged their employees to eat off washable plates in the office, rather than go out and grab food in containers that would end up in the landfill. To do this, they offered discounts on meals and engaged the cafeteria for food, rather than having pizza parties or catering that had too many landfill-only items.

Their fifth and final key insight was that systems don't rest. Whenever they thought they had the system figured out, there was a kink that was thrown in there. When they were at one of their highest waste diversion percentages, they hired a bunch of new people, which inevitably brought their waste diversion down. Whenever one problem was solved, another would seem to arise. It would be impossible to figure out these problems without systems thinking and the constant feedback that comes through feedback loops.[xxxv]

Chapter 10 - Practice Systems Thinking as an Individual

Improve Your Work

I've talked about different organizations practicing systems thinking, but you don't have to be the head of some large corporation to make it work for your life. In fact, anyone and everyone can benefit from using systems thinking one way or another throughout their personal life.

One way to practice systems thinking is to learn to ask different questions. When something goes wrong, we usually ask the same typical questions that never get us anywhere. Thinking in systems makes it easier to both find the bottleneck problems and improve or solve them. Start by paying more attention to the questions you ask.

Questions should be focused on the underlying relationships and patterns that are shown over time, and should focus your attention on how you and others perceive the different situations. What are the key patterns you should focus your questions on?

Let's take the case of business improvement. I will present you my case as a writer. I receive my royalties from Amazon with a 60-day delivery delay. This means what I made in January I will receive in March, and so on. Because the income flow is continuous, the hardest time for me was the first 60 days, because I got nothing for my hard work. Luckily, I had my nine-to-five job back then, so I didn't have to worry too much.

1.) Potential or actual delays in your system: If I had focused on those unpaid 60 days, I would probably have given up the entire writing business. Delays and oscillations still happen in my business, even after two years. For example, when I travel intensely and don't focus on researching and writing new books, my income decreases. However, this causes a severe perception and response delay from my side because even if I don't work in January and February, I won't feel the consequences because I receive my November and December royalties. Thus, the first month I will have that "oh my, oh my" moment is March.

I will start working like crazy in March, and thus in April, my royalties will start to increase again. However, my April royalties won't come into my account until June. This is the delivery delay my two lazy months have caused. Needless to say, it's frustrating to work 12 hours a day in March and April and still not see any results. But this is the system of self-publishing. I am familiar with the delays, so I can balance out the tight and abundant periods. This takes us to point two.

2.) Reinforcing processes: In my case, whenever I stop working it creates a reinforcing feedback loop in my business. The less I work, the fewer books I can publish, and the fewer books I publish, the less royalties I will receive. My old books are also "aging" and they bring less and less royalties each month. This, together with my general inactivity, exponentially decreases my income each month.

Whenever I start working again, I will create another reinforcing feedback loop (with long delays) and slowly start to crawl my income back to an acceptable level. And if I work even harder, it pays off even more.

3.) Unintended consequences: I need to accept that by taking one or two months off, I will cause unintended consequences in my business. Namely, I won't achieve an exponential growth, but rather a heavily oscillating one. For example, I will oscillate between $50 and $100 monthly income (for the example's sake) within a six-month period, but I won't go above $100 unless I make changes in the system—in my case, taking only two-week periods off, improving the durability of my books as best sellers, etc. This way, I can flatten the oscillations and eventually (when I have numerous books to sustain the $100 monthly income), I will be able to grow with each new book.

Exercise:

Using the three points of analysis above, try to construct a case for improving you work life.

Experience Time Differently

Another thing you can do is learn to experience time differently. Time is time, so how can you possibly do that? Well, when we are faced with hard-to-solve problems, we usually focus on short time intervals, much shorter than we should. When looking at how things are changing over time, try and redefine your present. Observing change through a longer block of time will actually help you make more accurate observations. For example, when I'm looking at the growth of my business in a six-month frame, I can see that I had $100 in January, $50 in April, and $100 in June. But if I looked at the two-year length, I would observe that a year ago in January, I had only $40. In the dip period, I had only $20, and then $40 again—but there still was a gradual, slow growth over time, and thus the next dip was actually still more than my peaking period a year before.

This being said, you can extend your present time you call "now" to mean a year ago and a year in the future—especially when you are analyzing the big picture of your business. It can help you more accurately determine what's happening and help you see relationships you hadn't seen before. What is happening now? What will the future hold for next year? If you see it all as a *now* moment, you'll be more eager to work even toward your one-year goal. We are creatures of instant gratification. If you make yourself believe that your future goal is happening in the present, you won't be procrastinating as much and you'll be more likely to work on it.

Information flow has sped up. We can find anything on the internet within a few seconds, and so we expect for everything else to be just as fast. Unfortunately, that doesn't happen. Peter Senge, the author of the book *The Fifth Discipline*, advises that slowing down and paying more attention to small details will help you. Calm yourself and appreciate life in the slow lane for once. Sit under a tree for half an hour or go for a walk without your phone. Observe the non-digital systems surrounding you. Notice what feedback loops the systems that surround you have. I'm not talking just about the forest and the trees, but rather, your personal life. There are still feedback loops which exist within the walls of your home and relationships. Maybe there is a reinforcing feedback loop of arguments because of some unwashed dishes. Maybe you can prevent an argument with others, maybe you can introduce something new into your home, or maybe you can just observe how the system is working. It doesn't hurt to know what you're dealing with.

If you want some more practice with systems learning, pick up your phone and check out the news. Read stories and try to see how patterns of behavior have changed over time in the current political or economic events you are following.

Collaborative Learning

There's nothing like getting together with others to collaboratively learn. Many people do this, and a lot of people get in what are called "mastermind groups." Individuals get together and talk, teach, learn, and problem-solve together. You've heard that two minds work better than one. When you invite another person to look at the things you are going through, you can see new perspectives that you never even considered. If you really want to learn, consider shadowing a coach or a mentor. This should be someone you admire and aspire to be like, and watching what they do throughout their day can help you see alternative ways people live.

Anyone else who is interested in systems thinking could be a good resource for you. Getting together in a group can help you sharpen and strengthen your skills. One idea is to have everyone bring a story from their professional life. You can then choose one to look at and work through to problem-solve and discuss as a group.

You can all read the most relevant books on the topic:

- *The Fifth Discipline* by Peter Senge
- *Thinking in Systems* by Donella Meadows
- *General Systems Theory* by Ludwig Von Bertalanffy
- *An Introduction to General Systems Thinking* by Gerald M. Weinberg
- *Redesigning the Future: Systems Approach to Societal Problems* by Russell L. Ackoff

The Society and You

Whether or not you are aware of it, you have an influence on society. Society also has an influence on you. Some of the ways you affect society and it affects you are through your significant relationships and the systems you participate in.

You could spend time analyzing your relationships and seeing what they are providing and what they are receiving from you. This can be done in a simple input and output model. Ask yourself:

- How do I influence the people in my relationships in this situation?
- How do they influence me?
- How do we influence our community?
- How do we influence our society?[xxxvi]

There are a lot of influences that happen here, and once you understand them, you can look at how to use your personal influence to make changes. This can improve both your life and the lives of those around you. I'm not talking about drastic changes. It's unlikely that you're going to go out and end poverty. It's simply not within your realm of influence. However, systems thinking makes it possible for you to change *something* that may lead to something bigger and bigger, and eventually make a countrywide or worldwide change. You have an influence on someone. They also have influences on others. And this begins a chain reaction. It's like lighting a match and watching the fire grow. One person really can make a big difference because of these relationships that exist within the system itself. Talking about certain situations and sharing ideas with others can help create change. As the momentum for these things gets rolling, more pressure is put on society for these changes to happen.

This is how revolutions start: someone starts talking about change to someone else. Then that other person talks to someone else, then many people will know about it and form a group. Their ideas will attract more and more people. The group grows into a movement. If the cause is important and strong enough, they can overthrow the entire system, just like in France's 18th-century revolution. But the ice bucket challenge gained momentum based on the same process.

This can be seen frequently in civil rights movements as well. Think about Martin Luther King Jr. People like him don't wake up one day and become this amazing, revolutionary public figure who everyone wants to watch and follow. It just doesn't happen like that. Instead, they start by influencing their friends and family. As that continues, their friends and family reach out to others, and then those individuals' friends and families reach out to more people. Pretty soon, you have a huge following of people who believe or think in similar ways. When all of these people rally together, the system has to change. The pressure influences society and new laws come out, changing the behavior of the system.

Exercise:

If you want to see how this works, grab a piece of paper. Right in the middle, write your name and a change you want to happen. It can be a big change or a small change. Then write down some of your strengths and skills. You can also write down things you don't necessarily like about yourself or things you want to change. Draw a circle around it, and this is now your self sphere.

Then draw a circle around that one and label it your family and friends. Write down how they influence you and draw an arrow back to the self sphere. Continue drawing circles and labeling the spheres with things like community, city, state, etc.

If you want to follow through with what you wrote down, think about how you can make this change. You'll need to jot down some daily, weekly, and monthly goals to make it happen. However, you should avoid making any goals or aspirations that are much too big. If you know for sure you won't achieve it, don't write it down. While you want to challenge yourself, you also want to be realistic. You'll only be motivated if the goal is realistic enough for you to meet.[xxxvii]

Personal Benefits

Have you ever seen a long line of people, but no one knows what they are waiting in line for? We tend to follow others, even if we cannot explain why we are doing it. It's easier to follow people who "know what they are doing" than to have to figure it out ourselves. We call this cognitive bias the bandwagon effect or herd effect.[xxxviii]

Your actions have consequences and you're responsible for your own behavior, even if you were following the crowd. Blindly following others could get you into some deep water, or at least put you in a situation you're not happy with. Assessing a situation from a systems thinking perspective, you will understand how the herd effect-action you are about to take will affect society and your own life. Just as Ben Stein said, "to getting the things you want out of life is this: Decide what *you* want." If you don't really know what you want yet from a systems point of view, at least you can analyze the possible outcome of possible actions you're about to take in the future. Systems thinking can help you choose the most advantageous path.

Conclusion

When you read through the book you might have been confused about what systems thinking was. You learned a bit more chapter by chapter. I recommend going through the book at least once more and read it with intention, now that you understand the basics.

Systems thinking is a relatively new discipline that can radically change how you see and address both complex and everyday issues. Instead of thinking in a linear and cause-effect way, you get to look at problems from a broader perspective. Almost everything you encounter is a system. That tree outside, the pond in your neighborhood, and even your relationships are systems. And systems are parts of other systems. It's like a Russian nesting doll where systems upon systems fit together.

Looking at systems, you can see the relationships and connections that determine how the system is functioning and behaving. These are complex problems, but you can actually see through them by mapping out the system. Take a look at the feedback loops in action, acknowledge the delays, and identify the unintended consequences a system produces. No longer do you have to suffer and try solution after solution, only to find that the problem is still there. Systems thinking helps you get rid of that Band-Aid approach and focus on solving the real problem instead.

Whether you're planning on using systems thinking within a large corporation or in your personal life, it can simplify how you live. It might seem more complex at first, but most systems thinkers never think in a linear way again once they truly understand how it works.

The important thing to remember is that systems thinking should not and cannot be rushed. It's a long learning process, and without going through every step, you won't get the desired results. This is especially true when trying to solve a complex problem in an organization or community.

Thank you for reading, and good luck in your future with systems thinking!

Yours,

Zoe

One more moment, please

How did you like Think in Systems? Would you consider leaving a feedback about your reading experience so other readers could know about it? If you are willing to sacrifice some of your time to do so, there are several options you can do it:

1. Please, leave a review on Amazon.
2. Please, leave a review on goodreads.com. Here is a link to my profile where you find all of my books.
   https://www.goodreads.com/author/show/14967542.Zoe_McKey
3. Send me a private message to zoemckey@gmail.com
4. Tell your friends and family about your reading experience.

Your feedback is very valuable to me to assess if I'm on the good path providing help to you and where do I need to improve. Your feedback is also valuable to other people as they can learn about my work and perhaps give an independent author as myself a chance. I deeply appreciate any kind of feedback you take time to provide me.

Thank you so much for choosing to read my book among the many out there. If you'd like to receive an update once I have a new book, you can subscribe to my newsletter at www.zoemckey.com. You'll get My Daily Routine Makeover cheat sheet and Unbreakable Confidence checklist for FREE. You'll also get occasional book recommendations from other authors I trust and know they deliver good quality books.

Brave Enough

Time to learn how to overcome the feeling of inferiority and achieve success. Brave Enough takes you step by step through the process of understanding the nature of your fears, overcome limiting beliefs and gain confidence with the help of studies, personal stories and actionable exercises at the end of each chapter.

Say goodbye to fear of rejection and inferiority complex once and for all.

Less Mess Less Stress

Don't compromise with your happiness. "Good enough" is not the life you deserve - you deserve the best, and the good news is that you can have it. Learn the surprising truth that it's not by doing more, but less with Less Mess Less Stress.

We know that we own too much, we say yes for too many engagements, and we stick to more than we should. Physical, mental and relationship clutter are daily burdens we have to deal with. Change your mindset and live a happier life with less.

Minimalist Budget

Minimalist Budget will help you to turn your bloated expenses into a well-toned budget, spending on exactly what you need and nothing else.

This book presents solutions for two major problems in our consumer society: (1) how to downsize your cravings without having to sacrifice the fun stuff, and (2) how to whip your finances into shape and follow a personalized budget.

Rewire Your Habits

Rewire Your Habits discusses which habits one should adopt to make changes in 5 life areas: self-improvement, relationships, money management, health, and free time. The book addresses every goal-setting, habit building challenge in these areas and breaks them down with simplicity and ease.

Tame Your Emotions

Tame Your Emotions is a collection of the most common and painful emotional insecurities and their antidotes. Even the most successful people have fears and self-sabotaging habits. But they also know how to use them to their advantage and keep their fears on a short leash. This is exactly what my book will teach you – using the tactics of experts and research-proven methods.

Emotions can't be eradicated. But they can be controlled.

The Art of Minimalism

The Art of Minimalism will present you 4 minimalist techniques, the bests from around the world, to give you a perspective on how to declutter your house, your mind, and your life in general. Learn how to let go of everything that is not important in your life and find methods that give you a peace of mind and happiness instead.

Keep balance at the edge of minimalism and consumerism.

The Critical Mind

If you want to become a critical, effective, and rational thinker instead of an irrational and snap-judging one, this book is for you. Critical thinking skills strengthen your decision making muscle, speed up your analysis and judgment, and help you spot errors easily.

The Critical Mind offers a thorough introduction to the rules and principles of critical thinking. You will find widely usable and situation-specific advice on how to critically approach your daily life, business, friendships, opinions, and even social media.

The Disciplined Mind

Where you end up in life is determined by a number of times you fall and get up, and how much pain and discomfort you can withstand along the way. The path to an extraordinary accomplishment and a life worth living is not innate talent, but focus, willpower, and disciplined action.

Maximize your brain power and keep in control of your thoughts.

In The Disciplined Mind, you will find unique lessons through which you will learn those essential steps and qualities that are needed to reach your goals easier and faster.

The Mind-Changing Habit of Journaling

Understand where your negative self-image, bad habits, and unhealthy thoughts come from. Know yourself to change yourself. Embrace the life-changing transformation potential of journaling.

This book shows you how to use the ultimate self-healing tool of journaling to find your own answers to your most pressing problems, discover your true self and lead a life of growth mindset.

The Unlimited Mind

This book collects all the tips, tricks and tactics of the most successful people to develop your inner smartness.

The Unlimited Mind will show you how to think smarter and find your inner genius. This book is a collection of research and scientific studies about better decision-making, fairer judgments, and intuition improvement. It takes a critical look at our everyday cognitive habits and points out small but serious mistakes that are easily correctable.

Who You Were Meant To Be

Discover the strengths of your personality and how to use them to make better life choices. In Who You Were Born To Be, you'll learn some of the most influential personality-related studies. Thanks to these studies you'll learn to capitalize on your strengths, and how you can you become the best version of yourself.

Wired For Confidence

Do you feel like you just aren't good enough? End this vicious thought cycle NOW. Wired For Confidence tells you the necessary steps to break out from the pits of low self-esteem, lowered expectations, and lack of assertiveness. Take the first step to creating the life you only dared to dream of.

To access the full list of my books visit <u>this link.</u>

Reference

Acaroglu, Leyla. Tools of a Systems Thinker. Medium. 2017. https://medium.com/disruptive-design/tools-for-systems-thinkers-the-6-fundamental-concepts-of-systems-thinking-379cdac3dc6a

Ackoff, Russel. Why Few Organizations Adopt Systems Thinking. The Systems Thinker. 2007. https://thesystemsthinker.com/why-few-organizations-adopt-systems-thinking/

Aronson, Daniel. Overview on Systems Thinking. Pegasus Communications (781) 398-9700. 1996.

AST Library. Systems Thinking for Social Change: Making an Explicit Choice (Book

Excerpt)" by David Peter Stroh. Applied Systems Thinking. 2018.
http://www.appliedsystemsthinking.com/supporting_documents/Leveraging_SocialChange.pdf

Balle, Michael. Managing With Systems Thinking: Making Dynamics Work for You in Business Decision Making. McGraw-Hill Book Co Lt. 1996.

Bertalanffy, Ludwig von. Untersuchungen über die Gesetzlichkeit des Wachstums. I. Allgemeine Grundlagen der Theorie; mathematische und physiologische Gesetzlichkeiten des Wachstums bei Wassertieren. Arch. Entwicklungsmech., 131:613-652. 1934.

Elmansy, Rafiq. The Six Systems Thinking Steps to Solve Complex Problems. Designorate. 2016.
http://www.designorate.com/systems-thinking-steps-solve-complex-problems/

Grubel, Ross. Systems Thinking: Connecting Self to Society. Corvis Group. 2015.
http://www.corvisgroup.com/systems-thinking-connecting-self-to-society/

Hoang, Jenny. Latimer, Leilani. A Tale of Three Bins. Sabre Inc. © and Presidio Graduate School. 2012.
https://www.sabre.com/images/uploads/A_Tale_of_Three_Bins_Sabre_Holdings_Case_Story.pdf

Ingraham, Christopher. Why hardly anyone dies from a drug overdose in Portugal. Washington Post. 2015.
https://www.washingtonpost.com/news/wonk/wp/2015/06/05/why-hardly-anyone-dies-from-a-drug-overdose-in-portugal/?noredirect=on&utm_term=.742c5c5da88c

Kendra, Cherry. What is the bandwagon effect? Very Well Mind. 2018.
https://www.verywellmind.com/what-is-the-bandwagon-effect-2795895

Kim, Daniel. Reinforcing And Balancing Loops: Building Blocks Of Dynamic Systems. Systems Thinker. 2018.
https://thesystemsthinker.com/reinforcing-and-balancing-loops-building-blocks-of-dynamic-systems/

Lisitsa, Ellie. The Four Horsemen: Criticism, Contempt, Defensiveness, and Stonewalling. The Gottman Institute. 2013.
https://www.gottman.com/blog/the-four-horsemen-recognizing-criticism-contempt-defensiveness-and-stonewalling/

Meadows, Donella. Thinking in Systems: A Primer. Chelsea Green Publishing. 2008.

Northwest Earth Institute. A systems thinking model: The Iceberg. Northwest Earth Institute. 2018. https://www.nwei.org/assets/A-SYSTEMS-THINKING-MODEL-The-Iceberg.pdf

Richmond, Barry. An Introduction to Systems Thinking. Isee systems, inc. 2004.

Ollhof, Jim. Walcheski Michael. Making The Jump to Systems Thinking. The Systems Thinker. 2018. https://thesystemsthinker.com/making-the-jump-to-systems-thinking/

Sparks, Chris. 104: Systems Thinking—The Essential Mental Models Needed for Growth. Medium. 2017. https://medium.com/@SparksRemarks/systems-thinking-the-essential-mental-models-needed-for-growth-5d3e7f93b420

Stroh, David Peter. Systems Thinking for Social Change. Chelsea Green Publishing. 2015.

Wayback Machine. His Life. Bertalanffy's Origins and his First Education. Wayback Machine. Retrieved 2018.
https://web.archive.org/web/20110725070225/http://www.bertalanffy.org/c_71.html

Weinberg, Gerald M. An Introduction to General Systems Thinking. Dorset House. 2001.

Endnotes

---

[i] Bertalanffy, Ludwig von. Untersuchungen über die Gesetzlichkeit des Wachstums. I. Allgemeine Grundlagen der Theorie; mathematische und physiologische Gesetzlichkeiten des Wachstums bei Wassertieren. Arch. Entwicklungsmech., 131:613-652. 1934.

[ii] Wayback Machine. His Life. Bertalanffy's Origins and his First Education. Wayback Machine. Retrieved 2018.

https://web.archive.org/web/20110725070225/http://www.bertalanffy.org/c_71.html

[iii] Richmond, Barry. An Introduction to Systems Thinking. Isee systems, inc. 2004.

[iv] Weinberg, Gerald M. An Introduction to General

Systems Thinking. Dorset House. 2001.

[v] Balle, Michael. Managing With Systems Thinking: Making Dynamics Work for You in Business Decision Making. McGraw-Hill Book Co Lt. 1996.

[vi] Meadows, Donella. Thinking in Systems: A Primer. Chelsea Green Publishing. 2008.

[vii] Acaroglu, Leyla. Tools of a Systems Thinker. Medium. 2017. https://medium.com/disruptive-design/tools-for-systems-thinkers-the-6-fundamental-concepts-of-systems-thinking-379cdac3dc6a

[viii] Northwest Earth Institute. A systems thinking model: The Iceberg. Northwest Earth Institute. 2018. https://www.nwei.org/assets/A-SYSTEMS-THINKING-MODEL-The-Iceberg.pdf

[ix] Picture IX. Acaroglu, Leyla. Tools of a Systems Thinker. Medium. 2017. https://medium.com/disruptive-design/tools-for-

systems-thinkers-the-6-fundamental-concepts-of-systems-thinking-379cdac3dc6a

[x] Meadows, Donella. Thinking in Systems: A Primer. Chelsea Green Publishing. 2008

[xi] Meadows, Donella. Thinking in Systems: A Primer. Chelsea Green Publishing. 2008

[xii] Meadows, Donella. Thinking in Systems: A Primer. Chelsea Green Publishing. 2008

[xiii] Meadows, Donella. Thinking in Systems: A Primer. Chelsea Green Publishing. 2008

[xiv] Picture XI. Meadows, Donella. Thinking in Systems: A Primer. Chelsea Green Publishing. Page 36. 2008

[xv] Kim, Daniel. Reinforcing And Balancing Loops: Building Blocks Of Dynamic Systems. Systems Thinker. 2018. https://thesystemsthinker.com/reinforcing-and-balancing-loops-building-blocks-of-dynamic-systems/

[xvi] Sparks, Chris. 104: Systems Thinking — The Essential Mental Models Needed for Growth. Medium. 2017.
https://medium.com/@SparksRemarks/systems-thinking-the-essential-mental-models-needed-for-growth-5d3e7f93b420

[xvii] Sparks, Chris. 104: Systems Thinking — The Essential Mental Models Needed for Growth. Medium. 2017.
https://medium.com/@SparksRemarks/systems-thinking-the-essential-mental-models-needed-for-growth-5d3e7f93b420

[xviii] Picture XIII. Sparks, Chris. 104: Systems Thinking — The Essential Mental Models Needed for Growth. Medium. 2017.
https://medium.com/@SparksRemarks/systems-thinking-the-essential-mental-models-needed-for-growth-5d3e7f93b420

[xix] Sparks, Chris. 104: Systems Thinking — The

Essential Mental Models Needed for Growth. Medium. 2017. https://medium.com/@SparksRemarks/systems-thinking-the-essential-mental-models-needed-for-growth-5d3e7f93b420

[xx] Ingraham, Christopher. Why hardly anyone dies from a drug overdose in Portugal. Washington Post. 2015. https://www.washingtonpost.com/news/wonk/wp/2015/06/05/why-hardly-anyone-dies-from-a-drug-overdose-in-portugal/?noredirect=on&utm_term=.742c5c5da88c

[xxi] Heart Of The Art. Systems Thinking: The Super Power of Autism. Heart Of The Art. 2018. https://www.heartoftheart.org/?p=4696

[xxii] Ackoff, Russel. Why Few Organizations Adopt Systems Thinking. The Systems Thinker. 2007. https://thesystemsthinker.com/why-few-

organizations-adopt-systems-thinking/

[xxiii] Ollhof, Jim. Walcheski Michael. Making The Jump to Systems Thinking. The Systems Thinker. 2018. https://thesystemsthinker.com/making-the-jump-to-systems-thinking/

[xxiv] Ollhof, Jim. Walcheski Michael. Making The Jump to Systems Thinking. The Systems Thinker. 2018. https://thesystemsthinker.com/making-the-jump-to-systems-thinking/

[xxv] Ollhof, Jim. Walcheski Michael. Making The Jump to Systems Thinking. The Systems Thinker. 2018. https://thesystemsthinker.com/making-the-jump-to-systems-thinking/

[xxvi] Picture XIV. Elmansy, Rafiq. The Six Systems Thinking Steps to Solve Complex Problems. Designorate. 2016. http://www.designorate.com/systems-thinking-steps-solve-complex-problems/

[xxvii] Aronson, Daniel. Overview on Systems

Thinking. Pegasus Communications (781) 398-9700. 1996.

[xxviii] Stroh, David Peter. Systems Thinking for Social Change. Chelsea Green Publishing. 2015.

[xxix] Stroh, David Peter. Systems Thinking for Social Change. Chelsea Green Publishing. 2015.

[xxx] AST Library. Systems Thinking for Social Change: Making an Explicit Choice (Book Excerpt)" by David Peter Stroh. Applied Systems Thinking. 2018.

http://www.appliedsystemsthinking.com/supporting_documents/Leveraging_SocialChange.pdf

[xxxi] AST Library. Systems Thinking for Social Change: Making an Explicit Choice (Book Excerpt)" by David Peter Stroh. Applied Systems Thinking. 2018.

http://www.appliedsystemsthinking.com/supporting_documents/Leveraging_SocialChange.pdf

[xxxii] Hoang, Jenny. Latimer, Leilani. A Tale of Three

Bins. Sabre Inc. © and Presidio Graduate School. 2012.

https://www.sabre.com/images/uploads/A_Tale_of_Three_Bins_Sabre_Holdings_Case_Story.pdf

[xxxiii] Picture VII. Hoang, Jenny. Latimer, Leilani. A Tale of Three Bins. Sabre Inc. © and Presidio Graduate School. 2012.

https://www.sabre.com/images/uploads/A_Tale_of_Three_Bins_Sabre_Holdings_Case_Story.pdf

[xxxiv] Picture VII. Hoang, Jenny. Latimer, Leilani. A Tale of Three Bins. Sabre Inc. © and Presidio Graduate School. 2012.

https://www.sabre.com/images/uploads/A_Tale_of_Three_Bins_Sabre_Holdings_Case_Story.pdf

[xxxv] Picture VII. Hoang, Jenny. Latimer, Leilani. A Tale of Three Bins. Sabre Inc. © and Presidio Graduate School. 2012.

https://www.sabre.com/images/uploads/A_Tale_of_Three_Bins_Sabre_Holdings_Case_Story.pdf

[xxxvi] Grubel, Ross. Systems Thinking: Connecting Self to Society. Corvis Group. 2015. http://www.corvisgroup.com/systems-thinking-connecting-self-to-society/

[xxxvii] Grubel, Ross. Systems Thinking: Connecting Self to Society. Corvis Group. 2015. http://www.corvisgroup.com/systems-thinking-connecting-self-to-society/

[xxxviii] Kendra, Cherry. What is the bandwagon effect? Very Well Mind. 2018. https://www.verywellmind.com/what-is-the-bandwagon-effect-2795895